Current Topics in Microbiology and Immunology

Volume 426

The review series *Current Topics in Microbiology and Immunology* provides a synthesis of the latest research findings in the areas of molecular immunology, bacteriology and virology. Each timely volume contains a wealth of information on the featured subject. This review series is designed to provide access to up-to-date, often previously unpublished information.

2019 Impact Factor: 3.095., 5-Year Impact Factor: 3.895

2019 Eigenfaktor Score: 0.00081, Article Influence Score: 1.363

2019 Cite Score: 6.0, SNIP: 1.023, h5-Index: 43

More information about this series at http://www.springer.com/series/82

Kenji Kabashima · Gyohei Egawa
Editors

Inducible Lymphoid Organs

Responsible Series Editor: Shizuo Akira

Springer

Editors
Kenji Kabashima
Department of Dermatology
Kyoto University
Kyoto, Japan

Gyohei Egawa
Department of Dermatology
Kyoto University
Kyoto, Japan

ISSN 0070-217X ISSN 2196-9965 (electronic)
Current Topics in Microbiology and Immunology
ISBN 978-3-030-51746-5 ISBN 978-3-030-51747-2 (eBook)
https://doi.org/10.1007/978-3-030-51747-2

This Springer imprint is published by the registered company Springer Nature Switzerland AG
The registered company address is: Gewerbestrasse 11, 6330 Cham, Switzerland

Preface

The antigen presentation to lymphocytes is one of the most important processes to develop adaptive immunity. We can functionally distinguish three distinct areas where lymphocytes recognize antigens:primary lymphoid organs (bone marrow and thymus), secondary lymphoid organs (lymph nodes and spleen), and the peripheral tissues. In the primary lymphoid organs, both T and B cells undergo positive and negative selections through recognizing self-antigens. In the secondary lymphoid organs, foreign antigens are presented by antigen-presenting cells, and antigen-specific lymphocytes are activated and expanded. Activated lymphocytes, especially T cells, circulate all over the body to survey foreign antigens, and once they recognize their cognate antigens, they promptly induce inflammation in an antigen-specific manner. Antigen recognition in the peripheral tissue is the final key process to develop adaptive immunity, thus to know about this process is essential for better control of immune-mediated diseases.

In the peripheral tissues, our body sometimes generates *de novo* lymphoid structure, called inducible lymphoid organs (ILOs). ILOs, also known as tertiary lymphoid organs or ectopic lymphoid organs, are organized as "sentinel" lymphoid tissues to be on the alert for invading pathogens, and are induced especially when the tissues are repeatedly exposed to foreign antigens, such as under chronic inflammation, infection, and cancers. ILOs have been observed in most tissues such as the intestine, lung, skin, genital tract, and exocrine glands. In each tissue, ILOs share many properties regarding its structure, cellular composition, and function, but also may have unique structures and functions in a context dependent manner.

In this book, readers will learn the basic structure and function of ILOs in the first chapter and will know the feature of ILOs in each tissue, which respectively have specific names, in the following chapters. We particularly highlights iBALT in the lung, M cells in the intestine, iSALT in the skin, MLCs in the genital tracts, and ILOs found in the synovial membrane in rheumatoid arthritis patients, and also covers the generations of artificially-constructed peripheral lymphoid tissues. Accordingly, readers will learn similarities and differences among ILOs in different sites. We invite the readers to visit recent findings showing how ILOs control adaptive immunity in each peripheral tissues.This book will offer fascinating and insightful contents for both scientists and clinicians in the areas of infectious and immune-associated diseases.

Kyoto, Japan
June 2020

Gyohei Egawa

Contents

Basics of Inducible Lymphoid Organs

Nancy H. Ruddle

Contents

Abstract Tertiary lymphoid organs (TLOs), also known as inducible lymphoid organs, tertiary lymphoid structures, tertiary lymphoid tissues, or ectopic lymphoid organs are accumulations of cells in chronic inflammation that have been observed in most tissues in autoimmunity, infection, and cancer in mouse and man. They share many properties with secondary lymphoid organs (SLOs), particularly lymph nodes, with regard to cellular composition, function, and regulation. TLOs include T and B cells, dendritic cells, follicular dendritic cells, and many other stromal cells, and high endothelial venules (HEVs) and lymphatic vessels. They serve as sites of antigen presentation and tolerance induction; they are harmful in autoimmunity and can be both harmful and beneficial in cancer. SLO induction in ontogeny is mediated by interactions of several cell types, including CD4+ CD3- lymphoid tissue inducer (LTi) RORγt+ cells that express LTαβ and interact with mesenchymal lymphoid tissue organizer (LTo) FAP+ cells in the presence of lymphatic

N. H. Ruddle (✉)
Epidemiology of Microbial Diseases, Yale School of Public Health, 60 College St., New Haven, CT 06510, USA
e-mail: Nancy.ruddle@yale.edu

Current Topics in Microbiology and Immunology (2020) 426: 1–20
https://doi.org/10.1007/82_2020_218

Published Online: 26 June 2020

and blood vessels. A variety of inducer cells initiate TLOs, including *bona fide* LTi cells, T cells, B cells, and NK cells. The mesenchymal organizer cells are less well characterized but can include FAP+ cells. Current challenges include identification of methods to inhibit TLOs in autoimmunity without affecting SLOs, and enhancement of TLOs for defense against tumors.

1 Introduction

Tertiary lymphoid organs (TLOs), also known as inducible lymphoid organs, ectopic lymphoid organs, tertiary lymphoid structures, or tertiary lymphoid tissues, are accumulations of lymphoid cells that arise in the course of chronic inflammation in infection (Neyt et al. 2012), autoimmunity, graft rejection, atherosclerosis, and cancer (Table 1). The term TLO was used in a review article in 1992 (Picker and Butcher 1992) to distinguish inflammation from classical lymphoid organs. Primary lymphoid organs (thymus, bone marrow, Bursa of Fabricius in birds) give rise in the course of ontogeny to the antigen recognizing cells of the immune system, namely T and B lymphocytes. Secondary lymphoid organs (SLOs), which also arise in ontogeny, include the lymph nodes (LNs), spleen, and mucosal-associated lymphoid tissue such as the nasal-associated tissue (NALT), tonsils, adenoids, and Peyer's patches, serve as site of antigen recognition and immune cell activation. TLOs are classified as lymphoid organs or tissues because of their cellular content, organization, and vasculature which resemble those of SLOs. However, they are *not* organs in that they do not have a defined immutable structure and location, and they generally do not possess a well-defined capsule (Table 2). TLO function, assumed to mimic that of secondary lymphoid organs, is a topic of intense speculation and research and potential therapeutic significance.

Information concerning the developmental regulation of SLOs has exploded in recent years. This provides a context for probing the origin and maintenance of TLOs that will provide crucial information for their experimental manipulation providing methodologies to limit them in pathology (e.g., autoimmunity) and to augment them in therapeutics (e.g., cancer).

2 Examples of Human and Murine TLOs

There were few mentions of TLOs after their initial description in 1992, probably because the field was not prepared for the concept that inflammation could develop into an organized accumulation of cells that resembled a lymph node. Inflammation at that time was defined by 4 clinical characteristics (heat, pain, redness, and swelling) rather than by the histological and functional characteristics that define different types and mechanisms of inflammation. The concept of TLOs received its impetus from the nearly simultaneous observations that mice deficient in

Table 1 Murine and Human Tertiary Lymphoid Organs

Disease	Affected organ	Species
Autoimmunity		
Hashimoto's thyroiditis	Thyroid	Mouse, human
Inflammatory bowel disease	Colon and small intestine	Mouse, human
Myasthenia gravis	Thymus	Mouse, human
Multiple sclerosis	Brain	Mouse, human
Rheumatoid arthritis	Joint	Mouse, human
Primary biliary cirrhosis	Liver	Human
Sjögren's syndrome	Lacrimal and salivary glands	Mouse, human
Systemic lupus erythematosus	Kidney	Mouse, human
Type 1 diabetes	Pancreas	Mouse, human?
Microbial Infection		
Borrelia burgdorferi	Joint	Human
Helicobacter pylori	Stomach	Human
Helicobacter spp.	Liver	Mouse
Influenza virus	Lung-iBALT	Mouse, human
Hepatitis virus	Liver	Human
Mycobacterium tuberculosis	Lung	Mouse human
Yersinia pseudotuberculosis	Adipose tissue (FALC)	Mouse
Chronic Graft Rejection	Heart, lung, kidney	Mouse, human
Atherosclerosis	Artery	Mouse, human
Cancer	Lung, colon, skin, prostate, ovary, pancreas, liver, testis, kidney, thyroid	Mouse, human

Legend: References for these conditions are found in the text and the following review articles [(Barone et al. 2016; Drayton et al. 2006; Neyt et al. 2012; Pipi et al. 2018)]. FALCs are described in (Han et al. 2017)

lymphotoxin-alpha (LTα) lacked lymph nodes (De Togni et al. 1994) and that rat insulin promoter (RIP) driven transgenic expression of LTα in the pancreas, kidney, and skin (RIPLTα mice) (Picarella et al. 1992) resulted in accumulations of cells at those sites that resembled lymph nodes (Kratz et al. 1996), the result of a process

Table 2 Comparison of Secondary and Tertiary Lymphoid Organs

Characteristics	Secondary lymphoid organ	Tertiary lymphoid organ
Origin	Ontogeny	Induced
Examples	Spleen, lymph node, Peyer's patches, tonsils, adenoids, MALT	Chronic Infection; autoimmunity; allograft rejection; cancer; iBALT; atherosclerosis
Location	Defined-relatively immutable	Within an organ or tissue
Lymphoid cells	T (CD4, CD8, Treg, Tfh), B, plasma cell	T (CD4, CD8, Treg, Tfh), B, plasma cell
Antigen presenting cells	DC, FDC, macrophage	DC, FDC, macrophage
Structures	B cell follicles, germinal centers, conduits	B cell follicles, germinal centers, conduits
Vessels	HEVs, lymphatic vessels	HEVs, lymphatic vessels
Stromal Cells	FRC, MRC	FRC
Capsule	Defined	Rare
Induction and maintenance	LTo (FAP+ podplanin+, LTβR+), LTi, endothelial cells, retinoic acid	LTo (FAP+ podoplanin+, LTβR+), ILC3/LTi, endothelial cells; CD4+ CD3+; NKT; Th17; γδT cells chronic inflammation; cytokines (LT, TNF, LIGHT, IL-17); chemokines (CXCL13, CCL19, CCL21)
Cytokines	LTαβ, LTα3	LTαβ, LTα3, TNF; IL-22; IL-17
Chemokines	CXCL13, CCL21, CCL19, CXCL12	CXCL13, CCL21, CCL19, CXCL12
Durability	Permanent (but can collapse)	Transient

MALT mucosal-associated lymphoid tissue; *iBALT* inducible bronchial-associated lymphoid tissue; *Treg* regulatory T cell; *Tfh* T follicular helper cell; *DC* dendritic cell; *FDC* follicular dendritic cell; *HEV* high endothelial venule; *FRC* fibroblast reticular cell; *MRC* marginal reticular cell; *LTo* lymphoid tissue organizer cell; *FAP* fibroblast activation protein-α; *LTi* lymphoid tissue inducer cell; FAP

termed lymphoid neo-organogenesis or neogenesis. This resulted in a revival of the concept of TLOs, although several different terms are now used to describe essentially the same functional and histological structures. There has been an explosion in the TLO literature with extensive documentation and several excellent review articles (Neyt et al. 2012; Barone et al. 2016; Colbeck et al. 2017; Drayton et al. 2006; Jones et al. 2016; Pipi et al. 2018) that list TLOs in man and mouse (Table 1). TLOs have several features in common that are noted below, that permit their designation as such, namely the lymphoid and stromal cellular content and organization characteristic of lymph nodes including the appropriate vascular compartments (high endothelial venules and lymphatic vessels). TLOs have been described in almost every organ, including, pancreas, lung, kidney, skin, brain, gut, liver, joints, artery, placenta, salivary and lacrimal glands, thymus, vagina, blood vessels, heart, testes, and ovaries.

Organized lymphoid cell accumulations in adipose tissue were noted in most species as early as 1863 and were termed milky spots (Von Recklinghausen 1863). More recently, they have been called fat-associated lymphoid clusters (FALC) (Elewa et al. 2014); they fit the definition of TLOs, with evidence of germinal centers (Benezech et al. 2015). These accumulations are present at a steady state, are reduced in germ-free mice suggesting a role for commensal organisms, and increase in the course of aging (Camell et al. 2019). They can be induced by zymosan injection (Benezech et al. 2015) or gram-negative bacterium *Yersinia pseudotuberculosis* infection (Han et al. 2017). Randolph and colleagues have described TLOs in Crohn's disease in the mesenteric fat that invade the lymphatic vessels that lead to LNs and suggested that they impede lymph flow in this condition (Randolph et al. 2016).

3 TLOs Share Structural and Cellular Features with LNs

LNs are bean-shaped structures found at defined locations in the body. They consist of a distinct capsule that contains collagen, smooth muscle cells, and lymphatic vessels (Fig. 1a). LNs are served by afferent lymphatic vessels that bring dendritic cells (DCs), soluble antigen, and lymphocytes. The lymph node is a mesh-like structure supported by a variety of stromal cells, including fibroblastic reticular cells (FRCs), follicular dendritic cells (FDCs), marginal reticular cells (MRCs), and CXCL12 abundant reticular cells (CRCs). A recent analysis by RNAseq identified as many as nine lymph node stromal cell clusters (Rodda et al. 2018). Some of the LN stromal cells are positive for the marker, fibroblast activation protein-α (FAP) (Denton et al. 2014) and FAP+ progenitors differentiate into many of these stromal cell populations (Denton et al. 2019). The LN is permeated by conduits, tubal structures lined by FRCs, that transport soluble antigens through the node. T cells and DCs are located in the paracortex, while B cells are located in the cortex with FDCs. Naïve lymphocytes enter the node through high endothelial venules (HEVs), specialized blood vessels that express ligands that facilitate the cells' recognition and entry. The HEVs of all immature LNs, Peyer's patches, and mature mesenteric nodes express mucosal addressin cell adhesion molecule-1 (MAdCAM-1), the ligand for a_4b_7 expressed on lymphocytes. Mature peripheral LN HEVs express CCL21 and several adhesion molecules including peripheral node addressin (PNAd), the ligand for CD62L (L-selectin) on lymphocytes. HEVs are tonically stimulated through the LTβR (Browning et al. 2005; Liao and Ruddle 2006) and are plastic in that they express different genes under conditions of homeostasis or inflammation (Liao and Ruddle 2006; Veerman et al. 2019). T cells are guided to the paracortical region through their CCR7 recognition of the chemokines CCL19 and CCL21 expressed by stromal cells and HEVs where they recognize antigen on DCs that have also been directed to the paracortical area through CCR7 interaction with CCL19 and CCL21. B cells are similarly directed to the cortical region via their expression of CXCR5 that allows migration to the chemokine, CXCL13, produced by stromal cells, including FDCs in the B cell follicles. B cells interact with antigen and cytokines produced by T follicular helper. (Tfh) cells in the follicles, giving rise to germinal centers, supported by the activity of activation-induced deaminanse (AID), resulting in plasmablast and plasma cell differentiation. Sphingosine-1-phosphate (S1P) is a lipid mediator that is found in high concentrations in the lymph and blood. Cells leaving the LN via efferent lymphatic vessels express high levels of an S1P receptor (S1P1) which interacts with its ligand, facilitating egress from the LN.

TLOs share many cellular and structural similarities with LNs, including their cellular composition-naïve and effector T cells, B cells, plasma cells, germinal centers, and antigen presenting cells, DCs, and FDCs (Fig. 1b) (Table 2). Their organization resembles that of LNs in that the cells are compartmentalized into distinct B and T cells zones due to the presence of the T zone chemokines, CCL19 and CCL21, and the B cell zone-defining chemokines, CXCL13 (Hjelmstrom et al. 2000) and CXCL12. Furthermore, most of the stromal characteristics of LNs have been noted in TLOs, namely conduits (Stranford and Ruddle 2012), HEVs (Kratz et al. 1996), lymphatic vessels (Kerjaschki et al. 2004), and FAP+ cells (Denton et al. 2019; Nayar et al. 2019). On the other hand, the well-defined capsule of LNs with its characteristic features, including a subcapsular sinus with antigen presenting cells, is rarely, if ever, seen. One characteristic that differentiates a TLO from a LN is that the former, even when it includes FAP+ and other typical LN stromal cells is located within another tissue, organ or tumor and thus is inescapably influenced by the microenvironment of that location, including cytokines, growth factors, and metabolites. Furthermore, in addition to microenvironmental differences between TLOs and LNs, there are environmental differences between TLOs in diverse locations. For example, the TLO in a tumor is subjected to very different influences than one in the pancreas in autoimmunity, including anoxia and suppressive cytokines such as IL-10 and TGFβ.

4 Activities and Functions of TLOs

What are the functions of TLOs? Why are they needed in the presence of an exquisite system that includes secondary lymphoid organs that provide sites for defense at strategic locations throughout the body? One suggestion is that TLOs represent primitive forms of lymphoid organs and that they arise in response to a chronic assault by a foreign pathogen when needed in infection. This is a logical explanation for the phenomenon—that is, that a TLO could arise at the site of a microbial infection, could sequester the foreign initiators of the infection, and, when the organisms have been quarantined from the remainder of the body (e.g., the chronic granuloma in tuberculosis), the TLO persists, or if organisms are eliminated, the TLO dissipates. Thus, one could think of TLOs as an early, local defense against microbes; they are apparent in the joint in Lyme arthritis (although by the time the TLO appears, the infectious agent, *Borrelia burgdorferi*, is no longer

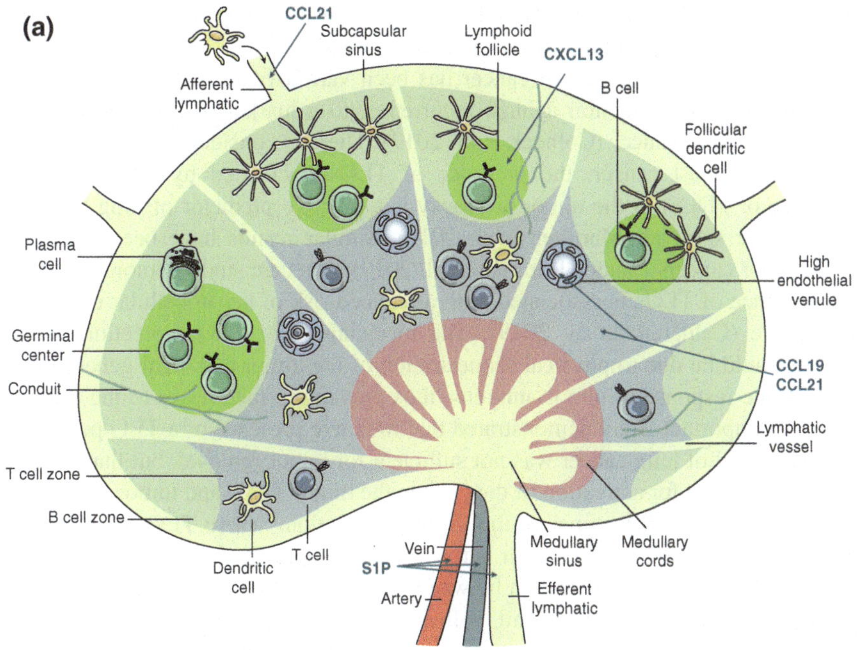

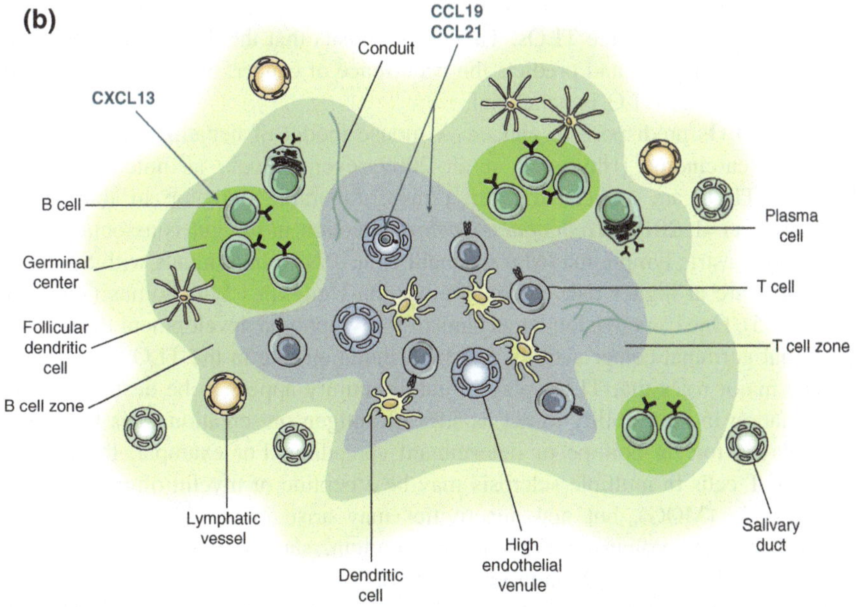

Fig. 1 Comparison of LN and TLO. A. Lymph node. B. TLO. A rendering of the appearance of a salivary gland in Sjögren's syndrome. (© Ruddle 2020)

detectable), in the lung in tuberculosis, in the liver in hepatitis B and C virus infection, and the gut in *Helicobacter pylori* infection.

The role that TLOs play in cancer has been variously attributed to a spectrum that ranges from protection against recurrence, to a marker of new cancers, to a predisposition to cancer (Colbeck et al. 2017). There is considerable evidence that, at least in breast cancer, the presence of TLOs (evaluated by HEV numbers, presence of FDCs) in the tumor at biopsy is a positive predictor of long term and disease-free survival (Martinet et al. 2011). Similar results have been found for many other cancers (Sautes-Fridman et al. 2019). A recent study demonstrated that the presence of TLOs is associated with a reduced risk of hepatocellular carcinoma recurrence (Calderaro et al. 2019). The mechanism for TLO protection against tumors could be due to physical sequestration of the tumor and local generation of an immune response to tumor antigens, thus preventing metastases to the draining LNs. An important study demonstrated that the mere presence of a TLO per se in a murine model of lung cancer was not sufficient for tumor removal, but that deletion of Tregs allowed for activation of the cells in the tumor TLOs and tumor destruction (Joshi et al. 2015). It is apparent that in those situations where TLOs contribute to defense, methods to enhance their activity or development should be pursued.

The close temporal association between some cancers and autoimmunity has led to the hypothesis by Rosen and colleagues that autoantigens released by the transformed cells activate the immune response (Rosen and Casciola-Rosen 2016). This phenomenon has been described in several instances of systemic autoimmunity (Shah et al. 2019, 2017), including rheumatoid arthritis and Sjögren's syndrome, which can include TLOs. This model posits that the TLO is a response to the tumor autoantigen and predicts the occurrence of cancer, but does not consider whether or not the TLO is beneficial.

Some TLOs predispose to cancer. A mouse model of hepatitis leads to hepatocellular carcinoma (Haybaeck et al. 2009) even though, as noted above, in humans, TLOs are associated with a more favorable outcome in that cancer (Calderaro et al. 2019). *Helicobacter pylori* infection in the gut is associated with TLOs and gastric lymphoma (Mazzucchelli et al. 1999). Individuals with Sjögren's syndrome are at higher risk of developing non-Hodgkin's lymphomas (Zintzaras et al. 2005). One can envision a mechanism for lymphoma development through the continual germinal center activity and high mutation rate in the TLO.

The major roles that TLOs play in autoimmunity appear to be destruction and perpetuation in their ability to act as sites of antigen presentation. It is likely that they play a role in epitope or determinant spreading. For example, the original target of T cells in multiple sclerosis may be a peptide of myelin oligodendrocyte glycoprotein (MOG), but new reactivities may arise in the TLO to additional peptides of that protein or even another protein, such as myelin basic protein (Kuerten et al. 2012; McMahon et al. 2005). Evidence that immunological reactivity in the TLO differs from that in the rest of the body and that the immune response actually occurs there comes from several examples. Mice transgenic for the rat insulin promoter driving LTα develop TLOs at the sites of transgene expression, the pancreas, kidney, and skin (Kratz et al. 1996). RIPLTα skin grafts that contain TLOs induce rejection of wild type skin allografts in mice that lack secondary lymphoid organs and even generate memory to alloantigens (Nasr et al. 2007). Vκ usage in the immunoglobulins of B cells of TLOs in rheumatoid arthritis differs from that of B cells in the circulation indicating that memory cells arise in the joint. These data support the concept that antibody affinity maturation is ongoing in the TLO, indicative of functional germinal centers at that location (Gause et al. 1997).

TLOs can also serve sites of tolerance induction or perpetuation. Their presence in long surviving heart (Baddoura et al. 2005) and kidney allografts (Brown et al. 2011) is suggestive of such a role. The presence of Tregs in many TLOs including those in atherosclerosis (Yin et al. 2016) and tumors (Joshi et al. 2015) suggests one possible mechanism of tolerance. Lymphatic endothelial cells (LECs) in lymph nodes are capable of self-antigen presentation and inducing tolerance (Cohen et al. 2014). It is likely that LECs in TLOs could play a similar role, and thus provide an additional mechanism of tolerance.

5 Regulation of TLOs

5.1 LN Development

The signals and cellular requirements for initiation of SLOs in ontogeny in the mouse have been fairly well elucidated and summarized in several reviews (Jeucken et al. 2019; Ruddle and Akirav 2009). The precise signals for the development of individual types of SLOs (peripheral LNs, mesenteric LNs, Peyer's patches, spleen) vary somewhat, but all depend on hematopoietic cells that express members of the LT/TNF family that interact with stromal cells expressing receptors for that family (Ruddle and Akirav 2009). LNs develop at defined locations throughout the body and require interactions between mesenchymal stromal cells, called lymphoid tissue organizer (LTo) cells and hematopoietic-derived lymphoid tissue inducer cells (LTi), which are a type of innate lymphoid cell (ILC3). The original 2-cell model has undergone some modification and requires consideration of the importance of blood and lymphatic endothelial cells that also play organizing roles through their expression of cytokines, cytokine receptors, chemokines, and adhesion molecules and their transport of lymphoid cells (Fig. 2a).

Recent data indicate that the embryonic and mature LTo cells are FAP+ (Denton et al. 2019). They express IL-7, LTβR, TNFR, VCAM, and MAdCAM-1. Retinoic acid derived from nerves has been reported to activate mesenchymal cells to produce CXCL13 (van de Pavert et al. 2009) in a location that is determined by LTβR on endothelial cells (Onder et al. 2017). Pre-LTi cells (CD45+ CD3- CD4+ Rorγt-) arrive via a vein between E14.5 and 15.0 at the inguinal LN anlagen that is already enmeshed in lymphatic vessels (Bovay et al. 2018). At later times (E15.0–18.5), additional LTi cells (CD45+ CD4+ CD3- a_4b_7+ RORγt+ IL7Rα+ LTα+

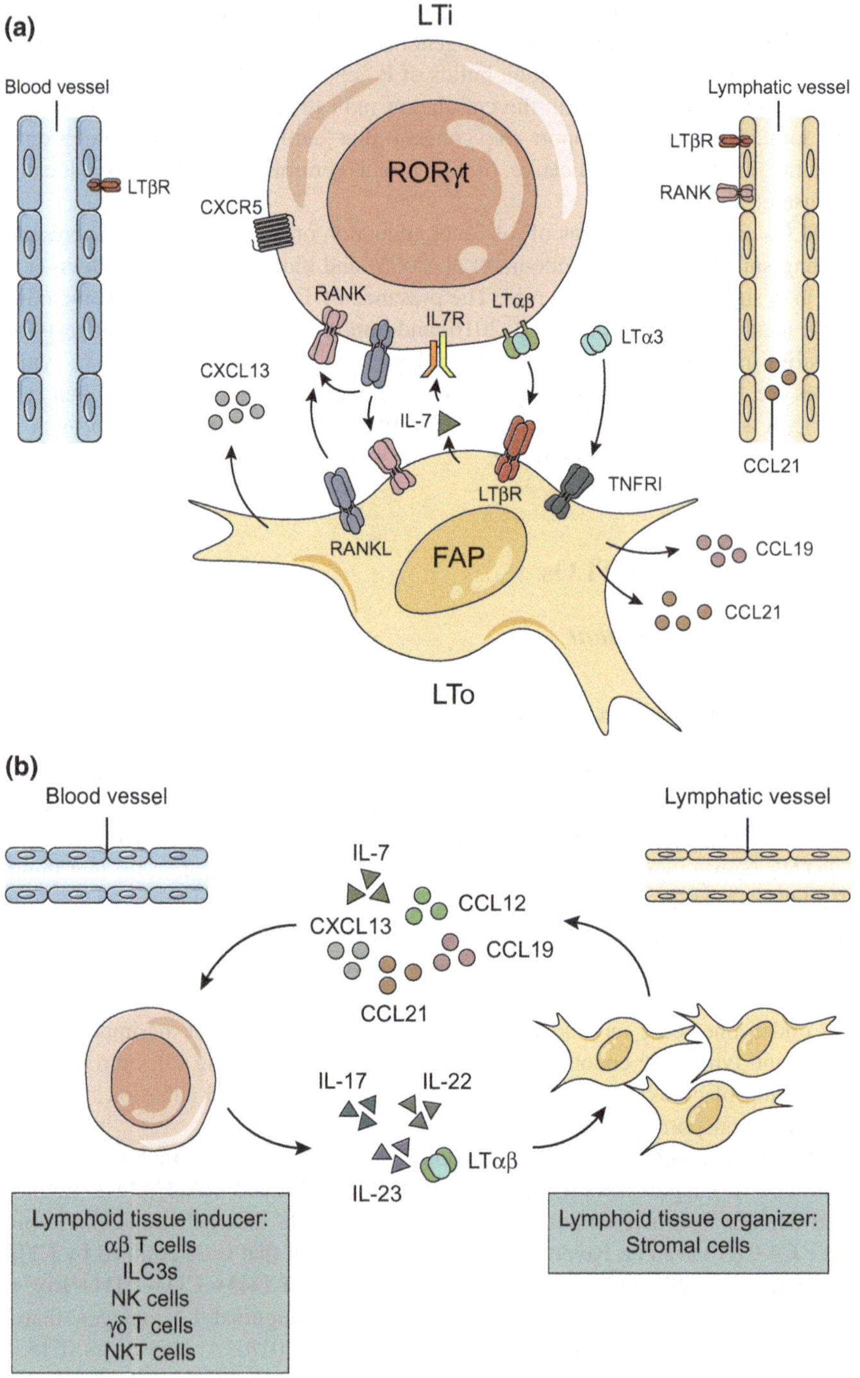

Fig. 2 Development of LN and TLO

LTβ+ CXCR5 + RANK+ RANKL+) are drawn into the region via expression of CXCL13, CCL19, CCL21, and IL7 by the LTo cells and enter into the LN anlagen via lymphatic vessels. Although lymphatic vessels clearly contribute in crucial ways to LN development, controversy exists concerning whether they participate *initiators* or as *facilitators* of the process (Koning and Mebius 2018; Onder and Ludewig 2018a, b), while recent studies have clearly defined the temporal relationship of lymphatic vessels (Bovay et al. 2018).

LTi cell expression of RANK and RANKL results in autocrine activation of LTα and further activation of LTo cells and lymphatic endothelial cells (Camara et al. 2019). CXCL13 produced by the LTo cells induces additional LTαβ from LTi cells. The LTo cells that express TNFRI and LTβR respond to LTα and LTαβ produced by the LTi cells and produce additional chemokines, including CCL19 and CCL21. Eventually, the lymphatic vessels contribute to formation of the LN capsule and subcapsular sinus. A recent publication suggests that even before birth a small population of blood vessels, which has some characteristics of HEVs, allows further accumulation of LTi cells and presumably B and T cells (Wang et al. 2018). Further production of LTα3 and LTα1β2 induces lymphoid chemokines from stromal cells and lymphatic vessels and partitioning of T cells and DCs in the paracortical regions and B cells and FDCs into the cortex. HEVs immediately after birth even in peripheral LNs express MAdCAM-1 and only at 24 h begin to express PNAd (Mebius et al. 1996). MAdCAM-1 is under the control of LTα3 and the TNFR1, whereas PNAd expression is regulated through the LTβR in part through LTαβ induction of Chst4 (Hemmerich et al. 2001) a crucial glucosyl transferase that modifies the scaffold proteins of PNAd.

5.2 *Induction of TLOs*

5.2.1 Central Questions

TLOs share many similarities with lymph nodes in regard to cellular composition and organization. Do they also arise through the same signals that induce LNs? Are LTo and LTi cells necessary for the development of TLOs? Or can other cells substitute? A key function of LTi cells in LN development is their production of LTαβ. However, in the adult, several cell types in addition to LTi cells produce the LTαβ complex, including T cells (CD4 TH1, Th17, CD8), B cells, and NK cells. The answer is that cells with all the characteristics of LTo and LTi cells can be found in TLOs, *but* other cells can take over the activities of those crucial embryonic players (Fig. 2b).

5.2.2 Stromal Cells

TLOs develop in non-lymphoid organs that have their own complement of stromal cells. Stromal cells in SLOs provide a scaffolding for cells and provide chemokines, growth factors, cytokines, and ligands for vascular addressins to allow the appropriate accumulation and organization of lymphoid cells. IL-7 appears to play an important role (Timmer et al. 2007) as it does in LN development. The sites of TLO development contain cells that assume the activities of LTos. As noted above, FAP+ cells are pre-LTo cells in LNs (Denton et al. 2019); these cells are found in TLOs in a mouse model of influenza virus, an infection in the lung. FAP+ immunofibroblasts are also seen in Sjögren's syndrome (SS) TLOs and in a mouse model of adenovirus-induced SS (Barone et al. 2016; Nayar et al. 2019). These FAP+ cells express chemokines (CXCL13, CCL19), podoplanin, ICAM-1 and VCAM-1, and IL-7 and thus have some of the characteristics of LTo cells. In this case, the cells are dependent for their activation on IL13 but independent of LTαβ and RORγt. One can envision other scenarios in other organs and under other TLO stimuli with dependence on LT. Lymphatic vessels are found in TLOs, as are HEVs that express MAdCAM-1 and/or PNAd, indicating that a source of cellular entrance and chemokine production are available as additional responders to LTi and LTi-like cells.

5.2.3 LTi Cells

Although LTi cells are crucial for lymphoid organ development in ontogeny, it is not intuitively obvious that they would persist in the adult once the lymphoid organs are formed. Furthermore, even if they do persist in the adult, what is their function in SLOs and TLOs? Cells with the characteristics of LTi cells are present in adult lymphoid organs (Kim et al. 2007) and are capable of restoring splenic architecture and function after infection with lymphocytic choriomeningitis virus (Scandella et al. 2008). CD4+ CD3- RORγt+ cells can restore lymphoid tissue to CXCR5-/- mice and are responsive to IL-7 (Schmutz et al. 2009). Furthermore, ILC precursors are found in peripheral blood in adult humans and mice; their expression of CD62L indicates that they could access HEVs and populate already existing SLOs and TLOs (Bar-Ephraim et al. 2019). However, this does not prove that they function as LTi cells in the initiation of TLOs. A subset of ILC3 cells that express neuropilin-1 (NRP-1) is found in human adult lymphoid tissue but not peripheral blood or skin (Shikhagaie et al. 2017). These NRP1+ cells express several markers of LTi cells (LTα, LTβ, RORγt) and can carry out activities in vitro consistent with those of LTi cells, including induction of adhesion markers on mesenchymal stromal cells. They are also found in lymphoid aggregates in the lungs of heavy smokers with chronic obstructive pulmonary disease, circumstantial evidence for their participation in induction of TLOs. LTi-like cells isolated from the lamina propria of adult mice also contain transcripts for NRP-1 (Robinette et al. 2015). The function, if any, of NRP-1 in the activity of LTi cells in development and in TLOs remains to be elucidated. Nevertheless, these data suggest that cells with the characteristics of LTi cells persist in adults, can exhibit functions of LTi cells in vitro and in vivo, and could conceivably participate in formation of TLOs.

5.2.4 LTi Cell Alternatives

Even if *bona fide* LTi cells are present in the adult, logic and experimental evidence indicate that though they may contribute to TLOs, they are not absolutely required for their maintenance and development. This was apparent from data obtained in the early transgenic experiments indicating that individual components of the LT/TNF family or downstream chemokines can induce TLOs in transgenic mice [reviewed in (Drayton et al. 2006)]. Even though fetal LTi cells are the sole producers of the various factors necessary for lymphoid organogenesis, several other cell types could play similar roles in the adult in their expression of the same factors and ability to induce the same chemokines and adhesion molecules. Notably, T cells, B cells, and NK cells all express LT (Ware et al. 1992).

When RORγt+ LTi cells are absent in adult mice, their function is assumed by other cell types. As noted above, LTi cells restore splenic architecture after viral infection, but mice that lack LTi cells still carry out this activity, albeit at a more leisurely pace (25 days compared to 16 days with RORγt+ intact cells) (Scandella et al. 2008). FALC can still develop in absence of LTi cells; in that case NKT cells take over their function (Benezech et al. 2015). Mice transgenic for thyroid expression of CCL21, a key lymphoid chemokine downstream of LT, still develop TLOs in the absence of LTi cells. This development is dependent on CD3+ CD4+ cells; their transfer, but not that of CD8 + or B220+ B cells, restores TLOs to RAG deficient CCL21+ tg mice (Marinkovic et al. 2006). The CD3+ CD4+ cells were clustered with DC cells in the nascent TLOs and their HEVs were inhibited by treatment with a LTβR-FC fusion protein. These data strongly support the conclusion that CD4 + cells (most likely expressing LTαβ), when appropriately stimulated by DCs, can support TLOs in the absence of LTi cells.

Th17 cells have been implicated in some TLOs. MOG-specific Th17 cell clones can transfer TLOs to syngeneic recipients (Jager et al. 2009) and disease severity is reduced in recipients deficient in the IL-17 receptor (Peters et al. 2011). Controversy exists concerning whether iBALT formation depends on IL-17. One report indicates that in the absence of LTi cells iBALT can be induced by CD4+ cells producing IL-17; another concludes that iBALT do not require IL-17. This controversy appears to be resolved by the realization that mechanisms and cytokine requirements for iBALT induction differ, depending upon the inducting agent. iBALT induced by LPS require IL-17 (Rangel-Moreno et al. 2011) as does iBALT induced by *Pseudonoas aeruginosa*. In the latter case iBALT, the source of IL-17 is γδ T cells rather than the conventional αβ Th17 cells (Fleige et al. 2011). On the other hand, iBALT induced by the poxvirus modified vaccinia virus Ankara does not require IL-17 (Fleige et al. 2011). IL-17 transgenic mice develop lung infiltrates but they are composed mainly of macrophages and do not meet the criteria of TLOs indicating that IL-17 alone, in contrast to LT and LIGHT, is not a primary inducer of TLOs. It is possible that additional cytokines produced by Th17 cells (LT, TNF, Il-22, IL-21) are necessary for the induction of TLOs.

5.3 TLO Plasticity

TLOs can be distinguished from LNs in their relative changeability and plasticity. This is not to say that LNs are completely static. On the contrary, they undergo extensive remodeling after immunization (Liao and Ruddle 2006), and, as noted above, the spleen undergoes drastic reorganization after viral infection (Scandella et al. 2008). However, these changes are brief and the organs usually return to the steady state except in the case of a devastating destruction of lymphocytes as in, e.g., AIDS. Even in that case, the structure of the LN remains. The case is quite different for TLOs. Changes in the nature of the infiltrate occur in the NOD mouse where Type 1 diabetes develops over the course of several weeks. The first histologic indication is an infiltration of lymphoid cells around and in the islet of Langerhans in the pancreas. The early (8 week) infiltrates are not organized and have the hallmarks of "inflammation" and lack HEVs. Later (20 weeks), the infiltrates have the appearance of TLOs, with T and B cell compartmentalization, HEVs, and lymphatic vessels. Even later, the lymphoid cells are activated, β cell destruction occurs, the mice are diabetic and the islet is fibrotic. The TLO has disappeared (Penaranda et al. 2010). A similar situation is seen when in RIPTNFα or RIPLTα mice when TNFα or LTα is expressed in the β cells in the pancreas of B6 mice (Picarella et al. 1993). There is a gradual organization of the cells into typical TLOs, which continue to enlarge over the age of the mouse. Treatment of the mouse with a variety of activators (super antigens, etc.) does little to change the appearance of the TLOs (A. Kratz PhD thesis). However, if RIPLTα or RIPTNFα mice are crossed with mice transgenic for B7-1 (CD80) in β cells, infiltrating cells become activated, the β cells are destroyed, the islet becomes fibrotic, and the mice are diabetic (Ruddle, NH unpublished) (Guerder et al. 1995).

6 Summary and Conclusions

In this chapter, the characteristics, regulation, and roles of TLOs have been described. These cellular infiltrates can be detrimental or beneficial, even in the same condition, as in cancer where they can predispose to tumors, or protect against them. The current challenge is to determine if there are any unique signals that would permit their manipulation; on the one hand, inhibiting them in autoimmunity, without seriously damaging the immune system and defense against pathogens, and, on the other hand, enhancing them in cancer. One possibility would be to target the LTαβ pathway since that is such a prominent feature of many TLOs.

Early studies indicated that treatment with an LTβR-IgG fusion protein resulted in reduction of HEVs and lymphoid chemokines in lacrimal glands and restoration of tear production in a mouse model of Sjögren's syndrome (Fava et al. 2011). However, a clinical trial of a similar compound (Baminercept) failed to alleviate clinical signs in Sjögren's syndrome human patients, despite reducing levels of plasma CXCL13 (St Clair et al. 2018). As noted above, it is likely that TLOs in the adult are the result of several different initiators and may not be as dependent upon the LTαβ signaling pathway as is SLO embryonic development. Thus, one could envision treating additional pathways that would induce TLOs in particular situations. The remarkable improvement in histological and clinical signs in some patients with RA after TNF inhibitor treatment suggests that pathway is a target for TLO amelioration. Current information suggests that treatment to inhibit TLOs while sparing SLOs and immune competence may be difficult because of the extensive similarities between the two (Table 2).

Targeting LT to tumors and subsequent development of TLOs was accomplished in a mouse model of melanoma by fusing the LTα gene to an antibody that recognizes a ganglioside on the tumor (Schrama et al. 2001). This resulted in development of TLOs at the tumor and reduction of the number of metastases. Of special interest was the presence of T cells in the tumor that recognized a melanocyte differentiation antigen. This treatment was effective even in mice that lacked all SLOs, providing evidence of antigen presentation and priming in the TLO at the tumor site (Schrama et al. 2008). These data indicate that directed expression of LT to tumors should result in TLOs capable of presenting antigen and eliminating cancer.

Conflict of Interest Dr. Ruddle is a consultant for L2 Diagnostics.

References

Baddoura FK, Nasr IW, Wrobel B et al (2005) Lymphoid neogenesis in murine cardiac allografts undergoing chronic rejection. Am J Transplant 5(3):510–516. https://doi.org/10.1111/j.1600-6143.2004.00714.x

Bar-Ephraim YE, Koning JJ, Burniol Ruiz E et al (2019) CD62L is a functional and phenotypic marker for circulating innate lymphoid cell precursors. J Immunol 202(1):171–182. https://doi.org/10.4049/jimmunol.1701153

Barone F, Gardner DH, Nayar S et al (2016) Stromal fibroblasts in tertiary lymphoid structures: a novel target in chronic inflammation. Front Immunol 7:477. https://doi.org/10.3389/fimmu.2016.00477

Benezech C, Luu NT, Walker JA et al (2015) Inflammation-induced formation of fat-associated lymphoid clusters. Nat Immunol 16(8):819–828. https://doi.org/10.1038/ni.3215

Bovay E, Sabine A, Prat-Luri B et al (2018) Multiple roles of lymphatic vessels in peripheral lymph node development. J Exp Med 215(11):2760–2777. https://doi.org/10.1084/jem.20180217

Brown K, Sacks SH, Wong W (2011) Tertiary lymphoid organs in renal allografts can be associated with donor-specific tolerance rather than rejection. Eur J Immunol 41(1):89–96. https://doi.org/10.1002/eji.201040759

Browning JL, Allaire N, Ngam-Ek A et al (2005) Lymphotoxin-beta receptor signaling is required for the homeostatic control of HEV differentiation and function. Immunity 23(5):539–550. https://doi.org/10.1016/j.immuni.2005.10.002

Calderaro J, Petitprez F, Becht E et al (2019) Intra-tumoral tertiary lymphoid structures are associated with a low risk of early recurrence of hepatocellular carcinoma. J Hepatol 70(1):58–65. https://doi.org/10.1016/j.jhep.2018.09.003

Camara A, Cordeiro OG, Alloush F et al (2019) Lymph node mesenchymal and endothelial stromal cells cooperate via the rank-rankl cytokine axis to shape the sinusoidal macrophage niche. Immunity 50(6):1467–1481 e1466. https://doi.org/10.1016/j.immuni.2019.05.008

Camell CD, Gunther P, Lee A et al (2019) Aging induces an NLRP3 inflammasome-dependent expansion of adipose B cells that impairs metabolic homeostasis. Cell Metab. https://doi.org/10.1016/j.cmet.2019.10.006

Cohen JN, Tewalt EF, Rouhani SJ et al (2014) Tolerogenic properties of lymphatic endothelial cells are controlled by the lymph node microenvironment. PLoS ONE 9(2):e87740. https://doi.org/10.1371/journal.pone.0087740

Colbeck EJ, Ager A, Gallimore A et al (2017) Tertiary lymphoid structures in cancer: drivers of antitumor immunity, immunosuppression, or bystander sentinels in disease? Front Immunol 8:1830. https://doi.org/10.3389/fimmu.2017.01830

De Togni P, Goellner J, Ruddle NH et al (1994) Abnormal development of peripheral lymphoid organs in mice deficient in lymphotoxin. Science 264(5159):703–707. https://doi.org/10.1126/science.8171322

Denton AE, Carr EJ, Magiera LP et al (2019) Embryonic FAP(+) lymphoid tissue organizer cells generate the reticular network of adult lymph nodes. J Exp Med. https://doi.org/10.1084/jem.20181705

Denton AE, Roberts EW, Linterman MA et al (2014) Fibroblastic reticular cells of the lymph node are required for retention of resting but not activated CD8+ T cells. Proc Natl Acad Sci U S A 111(33):12139–12144. https://doi.org/10.1073/pnas.1412910111

Drayton DL, Liao S, Mounzer RH et al (2006) Lymphoid organ development: from ontogeny to neogenesis. Nat Immunol 7(4):344–353. https://doi.org/10.1038/ni1330

Elewa YH, Ichii O, Otsuka S et al (2014) Characterization of mouse mediastinal fat-associated lymphoid clusters. Cell Tissue Res 357(3):731–741. https://doi.org/10.1007/s00441-014-1889-6

Fava RA, Kennedy SM, Wood SG et al (2011) Lymphotoxin-beta receptor blockade reduces CXCL13 in lacrimal glands and improves corneal integrity in the NOD model of Sjogren's syndrome. Arthritis Res Ther 13(6):R182. https://doi.org/10.1186/ar3507

Fleige H, Haas JD, Stahl FR et al (2011) Induction of BALT in the absence of IL-17. Nat Immunol 13(1):1 (author reply 2). https://doi.org/10.1038/ni.2167

Gause A, Gundlach K, Carbon G et al (1997) Analysis of VH gene rearrangements from synovial B cells of patients with rheumatoid arthritis reveals infiltration of the synovial membrane by memory B cells. Rheumatol Int 17(4):145–150. https://doi.org/10.1007/s002960050026

Guerder S, Carding SR, Flavell RA (1995) B7 costimulation is necessary for the activation of the lytic function in cytotoxic T lymphocyte precursors. J Immunol 155(11):5167–5174. Retrieved from https://www.ncbi.nlm.nih.gov/pubmed/7594526

Han SJ, Glatman Zaretsky A, Andrade-Oliveira V et al (2017) White adipose tissue is a reservoir for memory T cells and promotes protective memory responses to infection. Immunity 47(6):1154–1168 e1156. https://doi.org/10.1016/j.immuni.2017.11.009

Haybaeck J, Zeller N, Wolf MJ et al (2009) A lymphotoxin-driven pathway to hepatocellular carcinoma. Cancer Cell 16(4):295–308. https://doi.org/10.1016/j.ccr.2009.08.021

Hemmerich S, Bistrup A, Singer MS et al (2001) Sulfation of L-selectin ligands by an HEV-restricted sulfotransferase regulates lymphocyte homing to lymph nodes. Immunity 15(2):237–247. https://doi.org/10.1016/s1074-7613(01)00188-1

Hjelmstrom P, Fjell J, Nakagawa T et al (2000) Lymphoid tissue homing chemokines are expressed in chronic inflammation. Am J Pathol 156(4):1133–1138. https://doi.org/10.1016/S0002-9440(10)64981-4

Jager A, Dardalhon V, Sobel RA et al (2009) Th1, Th17, and Th9 effector cells induce experimental autoimmune encephalomyelitis with different pathological phenotypes. J Immunol 183(11):7169–7177. https://doi.org/10.4049/jimmunol.0901906
Jeucken CM, Konig JJ, Mebius RE et al (2019) The role of endothelial cells and TNF-receptor superfamily membes in lymphoid organogenesis and function during health and inflammation. Front Immunol 10:1–12. https://doi.org/10.3389/fimmun.2019.02700
Jones GW, Hill DG, Jones SA (2016) Understanding immune cells in tertiary lymphoid organ development: It is all starting to come together. Front Immunol 7:401. https://doi.org/10.3389/fimmu.2016.00401
Joshi NS, Akama-Garren EH, Lu Y et al (2015) Regulatory T cells in tumor-associated tertiary lymphoid structures suppress anti-tumor T cell responses. Immunity 43(3):579–590. https://doi.org/10.1016/j.immuni.2015.08.006
Kerjaschki D, Regele HM, Moosberger I et al (2004). Lymphatic neoangiogenesis in human kidney transplants is associated with immunologically active lymphocytic infiltrates. J Am Soc Nephrol 15(3):603–612. Retrieved from http://www.ncbi.nlm.nih.gov/pubmed/14978162
Kim MY, McConnell FM, Gaspal FM et al (2007) Function of CD4+ CD3—cells in relation to B- and T-zone stroma in spleen. Blood 109(4):1602–1610. https://doi.org/10.1182/blood-2006-04-018465
Koning JJ, Mebius RE (2018) Complexity of lymphoid tissue organizers: a response to Onder and Ludewig. Trends Immunol 39(12):951–952. https://doi.org/10.1016/j.it.2018.10.006
Kratz A, Campos-Neto A, Hanson MS et al. (1996) Chronic inflammation caused by lymphotoxin is lymphoid neogenesis. J Exp Med 183(4):1461–1472. Retrieved from http://www.ncbi.nlm.nih.gov/pubmed/8666904
Kuerten S, Schickel A, Kerkloh C et al (2012) Tertiary lymphoid organ development coincides with determinant spreading of the myelin-specific T cell response. Acta Neuropathol 124 (6):861–873. https://doi.org/10.1007/s00401-012-1023-3
Liao S, Ruddle NH (2006) Synchrony of high endothelial venules and lymphatic vessels revealed by immunization. J Immunol 177(5):3369–3379. https://doi.org/10.4049/jimmunol.177.5.3369
Marinkovic T, Garin A, Yokota Y et al (2006) Interaction of mature $CD3^+$ $CD4^+$ T cells with dendritic cells triggers the development of tertiary lymphoid structures in the thyroid. J Clin Invest 116(10):2622–2632. https://doi.org/10.1172/JCI28993
Martinet L, Garrido I, Filleron T et al (2011) Human solid tumors contain high endothelial venules: association with T- and B-lymphocyte infiltration and favorable prognosis in breast cancer. Cancer Res 71(17):5678–5687. https://doi.org/10.1158/0008-5472.CAN-11-0431
Mazzucchelli L, Blaser A, Kappeler A et al (1999) BCA-1 is highly expressed in Helicobacter pylori-induced mucosa-associated lymphoid tissue and gastric lymphoma. J Clin Invest 104 (10):R49–54. https://doi.org/10.1172/JCI7830
McMahon EJ, Bailey SL, Castenada CV et al (2005) Epitope spreading initiates in the CNS in two mouse models of multiple sclerosis. Nat Med 11(3):335–339. https://doi.org/10.1038/nm1202
Mebius RE, Streeter PR, Michie S et al (1996) A developmental switch in lymphocyte homing receptor and endothelial vascular addressin expression regulates lymphocyte homing and permits CD4+ CD3- cells to colonize lymph nodes. Proc Natl Acad Sci U S A 93(20):11019–11024. https://doi.org/10.1073/pnas.93.20.11019
Nasr IW, Reel M, Oberbarnscheidt MH et al (2007) Tertiary lymphoid tissues generate effector and memory T cells that lead to allograft rejection. Am J Transp 7(5):1071–1079. https://doi.org/10.1111/j.1600-6143.2007.01756.x
Nayar S, Campos J, Smith CG et al (2019) Immunofibroblasts are pivotal drivers of tertiary lymphoid structure formation and local pathology. Proc Natl Acad Sci U S A 116(27):13490–13497. https://doi.org/10.1073/pnas.1905301116
Neyt K, Perros F, GeurtsvanKessel CH et al (2012) Tertiary lymphoid organs in infection and autoimmunity. Trends Immunol 33(6):297–305
Onder L, Ludewig B (2018a) A fresh view on lymph node organogenesis. Trends Immunol 39 (10):775–787. https://doi.org/10.1016/j.it.2018.08.003

Onder L, Ludewig B (2018b) Redefining the nature of lymphoid tissue organizer cells: response to 'complexity of lymphoid tissue organizers' by Koning and Mebius. Trends Immunol 39 (12):952–953. https://doi.org/10.1016/j.it.2018.10.007

Onder L, Morbe U, Pikor N et al (2017) Lymphatic endothelial cells control initiation of lymph node organogenesis. Immunity 47(1):80–92 e84. https://doi.org/10.1016/j.immuni.2017.05.008

Penaranda C, Tang Q, Ruddle NH et al (2010) Prevention of diabetes by FTY720-mediated stabilization of peri-islet tertiary lymphoid organs. Diabetes 59(6):1461–1468. https://doi.org/10.2337/db09-1129

Peters A, Pitcher LA, Sullivan JM et al (2011) Th17 cells induce ectopic lymphoid follicles in central nervous system tissue inflammation. Immunity 35(6):986–996. https://doi.org/10.1016/j.immuni.2011.10.015

Picarella DE, Kratz A, Li CB et al (1992) Insulitis in transgenic mice expressing tumor necrosis factor beta (lymphotoxin) in the pancreas. Proc Natl Acad Sci U S A 89(21):10036–10040. Retrieved from http://www.ncbi.nlm.nih.gov/pubmed/1279667

Picarella DE, Kratz A, Li CB et al (1993) Transgenic tumor necrosis factor (TNF)-alpha production in pancreatic islets leads to insulitis, not diabetes. Distinct patterns of inflammation in TNF-alpha and TNF-beta transgenic mice. J Immunol 150(9):4136–4150. Retrieved from http://www.ncbi.nlm.nih.gov/pubmed/7682590

Picker LJ, Butcher EC (1992) Physiological and molecular mechanisms of lymphocyte homing. Annu Rev Immunol 10:561–591. https://doi.org/10.1146/annurev.iy.10.040192.003021

Pipi E, Nayar S, Gardner DH et al (2018) Tertiary lymphoid structures: autoimmunity goes local. Front Immunol 9:1952. https://doi.org/10.3389/fimmu.2018.01952

Randolph GJ, Bala S, Rahier JF et al (2016) Lymphoid aggregates remodel lymphatic collecting vessels that serve mesenteric lymph nodes in Crohn disease. Am J Pathol 186(12):3066–3073. https://doi.org/10.1016/j.ajpath.2016.07.026

Rangel-Moreno J, Carragher DM, de la Luz Garcia-Hernandez M et al (2011) The development of inducible bronchus-associated lymphoid tissue depends on IL-17. Nat Immunol 12(7):639–646. https://doi.org/10.1038/ni.2053

Robinette ML, Fuchs A, Cortez VS et al (2015) Transcriptional programs define molecular characteristics of innate lymphoid cell classes and subsets. Nat Immunol 16(3):306–317. https://doi.org/10.1038/ni.3094

Rodda LB, Lu E, Bennett ML et al (2018) Single-cell RNA sequencing of lymph node stromal cells reveals niche-associated heterogeneity. Immunity 48(5):1014–1028 e1016. https://doi.org/10.1016/j.immuni.2018.04.006

Rosen A, Casciola-Rosen L (2016) Autoantigens as partners in initiation and propagation of autoimmune rheumatic diseases. Annu Rev Immunol 34:395–420. https://doi.org/10.1146/annurev-immunol-032414-112205

Ruddle NH, Akirav EM (2009) Secondary lymphoid organs: responding to genetic and environmental cues in ontogeny and the immune response. J Immunol 183(4):2205–2212. https://doi.org/10.4049/jimmunol.0804324

Sautes-Fridman C, Petitprez F, Calderaro J et al (2019) Tertiary lymphoid structures in the era of cancer immunotherapy. Nat Rev Cancer 19(6):307–325. https://doi.org/10.1038/s41568-019-0144-6

Scandella E, Bolinger B, Lattmann E et al (2008) Restoration of lymphoid organ integrity through the interaction of lymphoid tissue-inducer cells with stroma of the T cell zone. Nat Immunol 9 (6):667–675. https://doi.org/10.1038/ni.1605

Schmutz S, Bosco N, Chappaz S et al (2009) Cutting edge: IL-7 regulates the peripheral pool of adult ROR gamma + lymphoid tissue inducer cells. J Immunol 183(4):2217–2221. https://doi.org/10.4049/jimmunol.0802911

Schrama D, Straten P, Fische WH et al (2001) Targeting of lymphotoxin-alpha to the tumor elicits an efficient immune response associated with induction of peripheral lymphoid-like tissue. Immunity 14(2):111–121. https://doi.org/10.1016/s1074-7613(01)00094-2

Schrama D, Voigt H, Eggert AO et al (2008) Immunological tumor destruction in a murine melanoma model by targeted LTalpha independent of secondary lymphoid tissue. Cancer Immunol Immunother 57(1):85–95. https://doi.org/10.1007/s00262-007-0352-x

Shah AA, Laiho M, Rosen A et al (2019) Protective effect against cancer of antibodies to the large subunits of both RNA polymerases I and III in Scleroderma. Arthritis Rheumatol 71(9):1571–1579. https://doi.org/10.1002/art.40893

Shah AA, Rosen A, Hummers LK et al (2017) Evaluation of cancer-associated myositis and scleroderma autoantibodies in breast cancer patients without rheumatic disease. Clin Exp Rheumatol 35 Suppl 106(4):71–74. Retrieved from https://www.ncbi.nlm.nih.gov/pubmed/28628466

Shikhagaie MM, Bjorklund AK, Mjosberg J et al (2017) Neuropilin-1 is expressed on lymphoid tissue residing LTi-like group 3 innate lymphoid cells and associated with ectopic lymphoid aggregates. Cell Rep 18(7):1761–1773. https://doi.org/10.1016/j.celrep.2017.01.063

St Clair EW, Baer AN, Wei C et al (2018) Clinical efficacy and safety of baminercept, a lymphotoxin beta receptor fusion protein, in primary Sjogren's syndrome: Results from a phase ii randomized, double-blind, placebo-controlled trial. Arthritis Rheumatol 70(9):1470–1480. https://doi.org/10.1002/art.40513

Stranford S, Ruddle NH (2012) Follicular dendritic cells, conduits, lymphatic vessels, and high endothelial venules in tertiary lymphoid organs: Parallels with lymph node stroma. Front Immunol 3:350. https://doi.org/10.3389/fimmu.2012.00350

Timmer TC, Baltus B, Vondenhoff M et al (2007) Inflammation and ectopic lymphoid structures in rheumatoid arthritis synovial tissues dissected by genomics technology: identification of the interleukin-7 signaling pathway in tissues with lymphoid neogenesis. Arthritis Rheum 56 (8):2492–2502. https://doi.org/10.1002/art.22748

van de Pavert SA, Olivier BJ, Goverse G et al (2009) Chemokine CXCL13 is essential for lymph node initiation and is induced by retinoic acid and neuronal stimulation. Nat Immunol 10 (11):1193–1199. https://doi.org/10.1038/ni.1789

Veerman K, Tardiveau C, Martins F et al (2019) Single-cell analysis reveals heterogeneity of high endothelial venules and different regulation of genes controlling lymphocyte entry to lymph nodes. Cell Rep 26(11):3116–3131 e3115. https://doi.org/10.1016/j.celrep.2019.02.042

Von Recklinghausen F (1863) Über eiter-bindegewebskörperchen. Virchows Arch Pathol Anat 28:157–166

Wang Z, Chai Q, Zhu M (2018) Differential roles of LTbetaRin endothelial cell subsets for lymph node organogenesis and maturation. J Immunol 201(1):69–76. https://doi.org/10.4049/jimmunol.1701080

Ware CF, Crowe PD, Grayson MH et al (1992) Expression of surface lymphotoxin and tumor necrosis factor on activated T, B, and natural killer cells. J Immunol 149(12):3881–3888. Retrieved from http://www.ncbi.nlm.nih.gov/pubmed/1281193

Yin C, Mohanta SK, Srikakulapu P et al (2016) Artery tertiary lymphoid organs: powerhouses of atherosclerosis immunity. Front Immunol 7:387. https://doi.org/10.3389/fimmu.2016.00387

Zintzaras E, Voulgarelis M, Moutsopoulos HM (2005) The risk of lymphoma development in autoimmune diseases: a meta-analysis. Arch Intern Med 165(20):2337–2344. https://doi.org/10.1001/archinte.165.20.2337

Role of iBALT in Respiratory Immunity

Aaron Silva-Sanchez and Troy D. Randall

Contents

Abstract Pulmonary respiration inevitably exposes the mucosal surface of the lung to potentially noxious stimuli, including pathogens, allergens, and particulates, each of which can trigger pulmonary damage and inflammation. As inflammation resolves, B and T lymphocytes often aggregate around large bronchi to form inducible Bronchus-Associated Lymphoid Tissue (iBALT). iBALT formation can be initiated by a diverse array of molecular pathways that converge on the activation and differentiation of chemokine-expressing stromal cells that serve as the scaffolding for iBALT and facilitate the recruitment, retention, and organization of leukocytes. Like conventional lymphoid organs, iBALT recruits naïve lymphocytes from the blood, exposes them to local antigens, in this case from the airways, and supports their activation and differentiation into effector cells. The activity of iBALT is demonstrably beneficial for the clearance of respiratory pathogens; however, it is less clear whether it dampens or exacerbates inflammatory responses to non-infectious agents. Here, we review the evidence regarding the role of iBALT in pulmonary immunity and propose that the final outcome depends on the context of the disease.

A. Silva-Sanchez · T. D. Randall (✉)
Department of Medicine, Division of Clinical Immunology and Rheumatology, University of Alabama at Birmingham, 1720 2nd AVE S, Birmingham, AL 35294, USA
e-mail: randallt@uab.edu

Current Topics in Microbiology and Immunology (2020) 426: 21–44
https://doi.org/10.1007/82_2019_191

Published Online: 24 January 2020

1 Introduction

Lymph nodes (LNs) are small, bean-shaped organs found along lymphatic vessels that drain the parenchyma of non-lymphoid organs. Like other secondary lymphoid organs (SLOs), LNs have a characteristic lymphoid architecture, with segregated B and T cell domains organized by distinct stromal cell types (Fletcher et al. 2011; Gentek and Bajenoff 2017). This structure facilitates the encounter of rare, antigen-specific lymphocytes with antigen-bearing dendritic cells (DCs) and thereby supports primary immune responses (Flajnik 2002; Neely and Flajnik 2016). LN formation occurs during late embryogenesis according to a developmental program that proceeds independently of antigen or inflammation (Luther et al. 2003). However, lymphocytes can also encounter antigen outside of LNs, as shown in reptiles and birds (species that lack LN), and in experimental mice that lack SLOs (Moyron-Quiroz et al. 2004). In these cases, T and B cells aggregate in the parenchyma of peripheral non-lymphoid organs and even form distinct B and T cell domains similar to those in conventional SLOs. Because these lymphoid aggregates do not occur as part of a developmental program and are only formed after local inflammation, they are termed tertiary lymphoid organs (TLO) (Hwang et al. 2016; Cupedo et al. 2004).

Three types of TLOs are found in the lung: nodular inflammatory foci (NIF), composed of clusters of myeloid cells and $CD8^+$ T cells (Stahl et al. 2013); granulomas, such as those formed during *Mycobacterium tuberculosis* infection, characterized by a central core of infected macrophages surrounded by B cells and T cells (Cadena et al. 2017); and inducible Bronchus-Associated Lymphoid Tissue (iBALT), which most closely resembles the architecture of conventional SLOs and is found in the perivascular space surrounding large blood vessels and along the airways of the lung (Hwang et al. 2016; Fleige and Forster 2017). iBALT formation occurs in response to numerous inflammatory conditions, using a variety of molecular pathways. In this chapter, we will summarize the current understanding of the steps leading to iBALT development and briefly review the impact of iBALT on pulmonary immune responses against microbial infections, allergens, and self-antigens.

2 Mechanisms Leading to iBALT Formation: A Rainbow of Options

The spatial distribution of lymphocytes in TLOs resembles that in SLOs, with the caveat that TLOs occur in places normally devoid of lymphocyte aggregates. iBALT typically forms on the basal side of the bronchial epithelium, often in the perivascular space of major blood vessels and consists minimally of a B cell follicle, sometimes with an active germinal center (GC) (Holt 1993). A variety of B cell phenotypes are observed in iBALT, including resting, naïve B cells, isotype-switched memory B cells, germinal center B cells, and antibody-secreting plasma cells (GeurtsvanKessel et al. 2009; Halle et al. 2009; Rangel-Moreno et al. 2006). T cells and DCs are located along the bronchial epithelium and typically surround the B cell follicle (Fig. 1a) (Halle et al. 2009).

The organization, maintenance, and survival of leukocytes in iBALT require the presence of specialized stromal cells. For example, CD31$^+$PNAd$^+$ high endothelial venules (HEVs) form near the outer edges of the B cell follicle and serve as entry portals for recirculating lymphocytes (Ager 2017; Otsuki et al. 1989; Sato et al. 2000). Newly formed Thy1$^+$ lymphatic endothelial cells (LECs) appear in the lungs after an inflammatory response, particularly surrounding areas of iBALT, where they support T cell recruitment and survival by secreting the chemokines, CCL21 and CCL19, as well as the cytokines, IL-7 and IL-33 (Baluk et al. 2009, 2014a). In SLOs, the formation of B cell follicles depends on the secretion of CXCL13 by a network of follicular dendritic cells (FDCs) that attract CXCR5$^+$ B cells (Carlsen et al. 2002; Yu et al. 2002). However, two types of B cell follicles are described in iBALT—a classic follicle with CD35$^+$CXCL13$^+$ FDCs (Rangel-Moreno et al. 2011) and non-classical B cell follicle that lacks FDCs and instead uses podoplanin (PDPN)$^+$CD35$^-$CD31$^-$CXCL12$^+$ fibroblast-like stromal cells to maintain the B cell

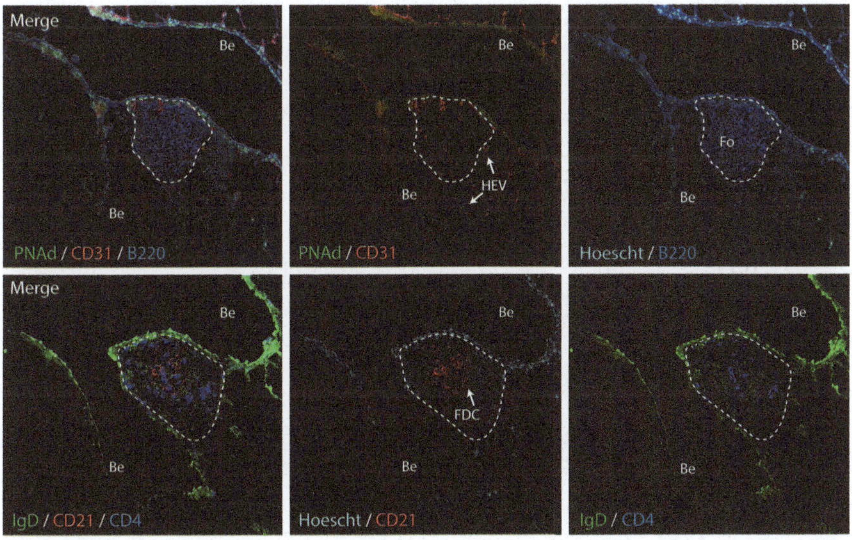

Fig. 1 Structure of iBALT. Immunofluorescence staining was performed on serial sections of lungs from 3-week-old mice after treatment with LPS during the first week after birth. Top row shows B cell follicle in iBALT (B220$^+$ cells in blue delimited by the dotted line) and associated HEV structures (PNAd$^+$ in green and CD31$^+$ n red) indicated by arrows. CD31$^+$ blood vasculature is observed in the whole field. Bronchi epithelia (Be) are indicated with arrowheads. The bottom row shows IgD$^+$ B cells inside the B cell follicle of iBALT (IgD$^+$ in green) and the associated FDC network that supports B cell aggregation (CD35$^+$ cells in red). Scattered CD4 T cells can be seen within and near the outer edges of the B cell follicle Be indicates bronchial epithelium

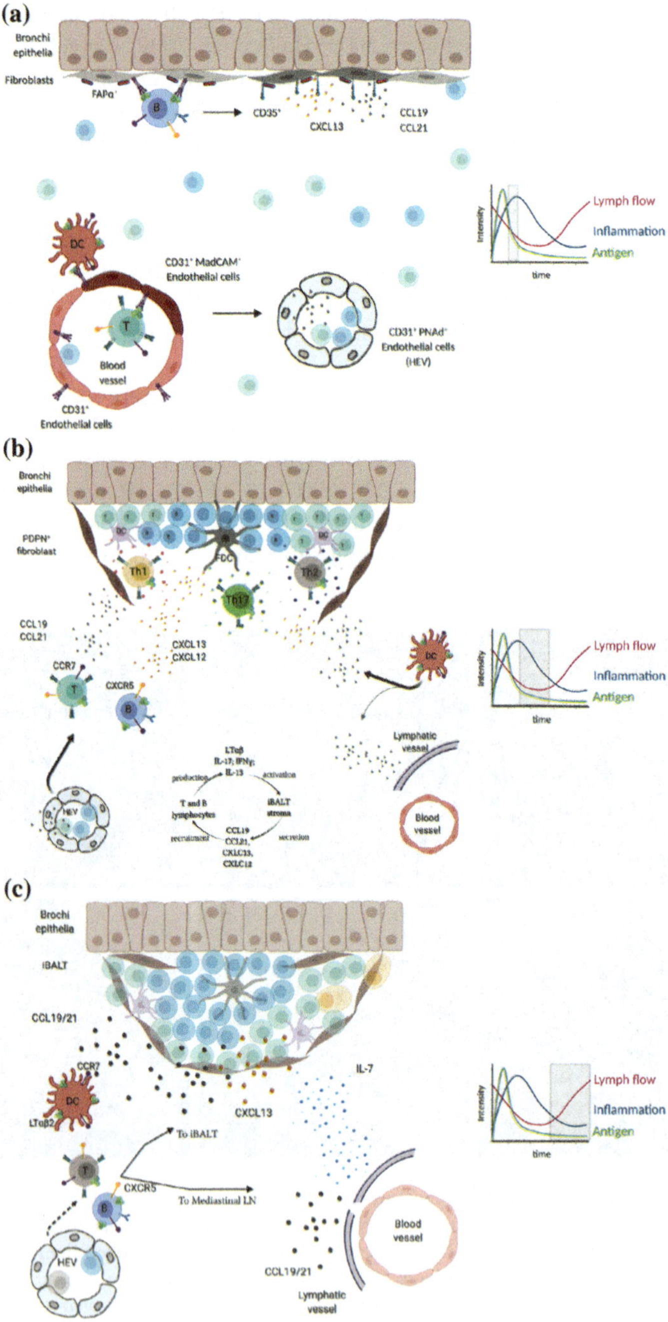
(a)
Bronchi epithelia
Fibroblasts
FAPα⁺
B
CD35⁺
CXCL13
CCL19
CCL21
DC
CD31⁺ MadCAM⁺ Endothelial cells
T
Blood vessel
CD31⁺ Endothelial cells
CD31⁺ PNAd⁺ Endothelial cells (HEV)
Intensity
time
Lymph flow
Inflammation
Antigen
(b)
Bronchi epithelia
PDPN⁺ fibroblast
DC
FDC
Th1
Th2
Th17
CCL19
CCL21
CXCL13
CXCL12
CCR7
T
CXCR5
B
DC
Lymphatic vessel
Blood vessel
HEV
LTαβ
IL-17; IFNγ;
IL-13
production
activation
T and B lymphocytes
iBALT stroma
CCL19
CCL21,
CXLC13,
CXLC12
recruitment
secretion
Intensity
time
Lymph flow
Inflammation
Antigen
(c)
Brochi epithelia
iBALT
CCL19/21
CCR7
DC
LTαβ2
IL-7
CXCL13
To iBALT
T
CXCR5
B
To Mediastinal LN
HEV
Blood vessel
CCL19/21
Lymphatic vessel
Intensity
time
Lymph flow
Inflammation
Antigen

◂**Fig. 2 Sequential development of iBALT**. **a** Early stage of iBALT formation requires the differentiation and activation of stromal cells that segregate B cells and T cells. CXCL13 producing FDCs arise from the activation of FAPα+ fibroblastic cells in a LTαβ dependent manner. **b** Local inflammation is required to amplify the activation of newly developed stromal cells and the recruitment of activated lymphocytes and dendritic cells. During IAV infection model, this stage would correspond to the lymphocyte aggregates observed right after clearance of the virus around 14 days post-infection. **c** iBALT structures perdure after the peak of the inflammatory response. At this stage, the feedback loop between stromal cells and the recruitment of LTαβ-bearing lymphocytes is controlled by the frequency of recruitable $LT\alpha\beta^+$ lymphocytes

area (Fleige et al. 2014). Recruitment of B cells toward the $PDPN^+CXCL12^+$ cells requires the expression of CXCR4 by B cells, similar to that described in the dark zones of germinal centers in conventional SLOs (Rodda et al. 2015). While it remains to be elucidated whether the two types of B cell follicles in iBALT are functionally different, the differentiation of $CXCL12^+PDPN^+$ stroma requires IL-17 signaling (Fig. 1b).

A wide range of stimuli trigger iBALT formation including viruses (Moyron-Quiroz et al. 2004; GeurtsvanKessel et al. 2009; Rangel-Moreno et al. 2007), bacteria (Baluk et al. 2014a; Fleige et al. 2014), fungi (Eddens et al. 2017), helminths (Venturiello et al. 2007; Gentilini et al. 2011), microbial products, particulates, and other inflammatory stimuli (Kuroda et al. 2016; Gregson et al. 1979; Noble and Zhao 2016). iBALT formation also occurs in mice that overexpress (Botelho et al. 2013; Furtado et al. 2014) or lack (Kocks et al. 2007; Das et al. 2006; Bouton et al. 2012) particular genes, each of which gives us insight into the mechanisms that lead to TLO formation. In general, the development of iBALT parallels the embryonic development of SLOs (Fig. 2). The first step of iBALT development entails the activation and differentiation of stromal cell precursors into iBALT supporting stroma—FDCs, HEVs, LECs, and $CXCL12^+PDPN^+$ fibroblasts (Fig. 2a). In the second step, leukocyte-mediated inflammation increases the recruitment of B and T cells around the activated stroma, leading to the maturation of the iBALT structure (Fig. 2b). In the third step, inflammation resolves and iBALT is maintained by homeostatic interactions between leukocytes and stromal cells, which supports the recruitment and organization of leukocytes from the blood (Fig. 2c).

The differentiation of each stromal cell type uses distinct molecular pathways (Lu and Browning 2014; Girard et al. 2012). For example, the formation of LECs depends on the secretion of IL-1β, which leads to VEGF expression and signaling through VEGFR2 and VEGFR3 (Baluk et al. 2009, 2013, 2014a, b). Although the formation of iBALT and the appearance of new lymphatic vessels occur at the same time and are closely associated, the formation of new lymphatic vessels is independent of iBALT development (Baluk et al. 2014a). The role of LTβR signaling is not clear since LTβ-deficient mice develop more lymphatic vessels than LTα-deficient and WT mice when infected with *Mycoplasma pulmonis* (Mounzer et al. 2010). However, the ectopic expression of LTα promotes TLO development (Mounzer et al. 2010), including the formation of new lymphatic vessels, suggesting that LTα_3 is sufficient to trigger the generation of new lymphatic vessels and TLO. In contrast, the differentiation of FDCs, PDPN$^+$ fibroblasts, and HEVs heavily depends on LTβR signaling (Lu and Browning 2014), although in some cases (i.e., LTαβ-deficient mice), it can be bypassed by the overexpression of TNFα (Furtado et al. 2014; Guedj et al. 2014), suggesting that other members of the TNF superfamily can act as triggers of FDC and PDPN$^+$ fibroblast differentiation (Ciccia et al. 2017; Ding et al. 2016; Berrih-Aknin et al. 2013).

A common mesenchymal stromal cell precursor in the LN gives rise to marginal reticular cells (MRC), fibroblastic reticular cells (FRC), some FDCs, and CRCs, but not HEVs (Denton et al. 2019a). This mesenchymal stromal cell precursor expresses the fibroblast activation protein alpha (FAPα), VCAM, CXCL13, and LTβR and is found in the perivascular region as lymphoid tissue inducer cells are being recruited to the LN anlagen (Denton et al. 2019a). A similar fibroblastic FAPα$^+$ cell is found in the lungs of mice infected with influenza, one of the stimuli that promotes iBALT formation (GeurtsvanKessel et al. 2009; Denton et al. 2019a, b). Fate-mapping shows that FAPα$^+$ mesenchymal cells are the precursors of CD35$^+$FAPα$^+$ FDC-like cells in the B cell follicles of iBALT (Denton et al. 2019a). Unlike in the LN anlagen, however, the differentiation of FAPα$^+$ precursors into mature stromal cells does not require the subset of innate lymphoid cells known as lymphoid tissue inducer (LTi) cells (Denton et al. 2019a), most likely because of numerous cell types, including activated B cells, T cells and DCs, can express LT and promote stromal cell maturation (Rangel-Moreno et al. 2011; Furtado et al. 2014; Marinkovic et al. 2006).

In conventional SLOs, the transition of CD31$^+$MadCAM$^-$PNAd$^-$ blood endothelial cells (BECs) to CD31$^+$MadCAM$^+$PNAd$^-$ immature HEVs and to CD31$^+$MadCAM$^-$PNAd$^+$ mature HEVs requires signaling through the LTβR and the activation of the canonical (RelA, p52) and non-canonical (RelB) NF-kB pathways (Ager 2017). In adult mice, LN HEVs need to be maintained by the constant influx of homeostatically activated DCs arriving from the afferent lymphatics (Herzog et al. 2013; Baratin et al. 2015; Astarita et al. 2015). In fact, any interruption of lymph flow (Mebius et al. 1991a, b), DC influx (Moussion and Girard 2011; Wendland et al. 2011), or LTβR/NF-kB signaling (Martinet et al. 2013; Browning et al. 2005) leads to the rapid involution of HEVs into flattened endothelial cells that lack PNAd expression. Similar pathways regulate iBALT-associated HEVs that surround the B cell follicle (Weinstein and Storkus 2016; Sato et al. 2011; Drayton et al. 2003).

Interestingly, a wide variety of signals can trigger the initial differentiation of stromal cells and start the process of iBALT formation. For example, the administration of LPS to neonatal mice leads to a strong IL-17 response, which turns on CXCL13 and promotes iBALT formation (Rangel-Moreno et al. 2011). In fact, IL-17 seems to be involved in the formation of iBALT and other TLOs in numerous contexts (Rangel-Moreno et al. 2011; Fleige et al. 2012, 2014; Eddens et al. 2017). Another Th17-related cytokine, IL-22, which is involved in epithelial repair and TLO formation in other tissues (Aujla and Kolls 2009; Barone et al. 2015; Pociask et al. 2013; Rendon et al. 2013), may also play a role in iBALT formation, as B cell follicles are mildly reduced in size and number in the lungs of *M. tuberculosis*-infected IL-22-deficient mice (Khader et al. 2011).

Th17-related molecules are not the only inducers of iBALT formation, as mice infected with modified Vaccinia Ankara develop iBALT in an IL-17-independent fashion (Fleige et al. 2012, 2014). Mice infected with influenza also develop iBALT. In this case, however, type I IFN signaling is responsible for CXCL13 expression by lung fibroblasts and subsequent formation of B cell aggregates (Denton et al. 2019b). Similarly, mice infected with *Pneumocystis murina* develop iBALT in response to a mixed Th2 (IL-13) and Th17 (IL-17A) response, in which CXCL13 expression by lung $PDPN^+$ fibroblasts is dependent on the synergistic effects of IL-13 and IL-17 on IL-6 (Eddens et al. 2017). The pulmonary administration of particulates like alum triggers iBALT formation via macrophage cell death and IL-1α release (Kuroda et al. 2016). Similarly, the IL-1-related cytokines, IL-36 and IL-18, promote the formation of TLOs in colorectal cancer (Weinstein et al. 2019) and iBALT formation in COPD patients (Briend et al. 2017). Taken together, these data suggest that the first step of iBALT development depends on the differentiation of stromal cells capable of recruiting and organizing leukocytes via the production of chemokines like CXCL13 and that subsequently, the accumulation of activated, LT-expressing lymphocytes generate a positive feedback loop that maintains the structure.

Interestingly, the impairment of lymphatic drainage from the lungs is sufficient to trigger the formation of iBALT (Reed et al. 2019). Because the lymphatic vessels in the lungs of humans and mice lack smooth muscle cells in the lymphangions responsible for collecting lymph (Reed et al. 2019), the lymph flow from the lung depends on changes in the thoracic pressure produced by respiration. However, mice with a platelet-specific deletion of CLEC2, a ligand for PDPN and highly expressed on platelets, have impaired lymphatic flow from the lungs and spontaneously develop iBALT (Reed et al. 2019). Moreover, the ablation of $CD11c^+$ cells, presumably DCs, leads to the dissolution of iBALT structures (GeurtsvanKessel et al. 2009). Although $CD11c^+$ cell depletion affects cells other than DCs, including activated B cells (Zhang et al. 2019; Winslow et al. 2017; Naradikian et al. 2016) and some FDCs (Aziz et al. 1997), these studies suggest that DCs are important for the homeostatic maintenance of iBALT. Consistent with this idea, the loss of CCR7 on $CD11c^+$ cells leads to iBALT formation (Halle et al. 2009; Fleige et al. 2018), perhaps because activated DCs accumulate in the lung. Together these data suggest that DCs help maintain iBALT by providing LT signals to stromal cells (Muniz et al. 2011). Interestingly, CCR7 also regulates the trafficking of regulatory $CD4^+$ T cells (Tregs), which are important for limiting inflammatory responses (Georgiev et al. 2019). Neonatal mice lacking CCR7 spontaneously form iBALT due to impaired Treg migration and loss of inflammatory control (Foo et al. 2015; Cowan et al. 2013). However, once iBALT is formed, it recruits $FoxP3^+$ Tregs (Li et al. 2019; Trujillo et al. 2010; Siemeni et al. 2019), which help limit local inflammatory responses. These data indicate that once formed, iBALT is maintained by homeostatic mechanisms similar to those that maintain conventional SLOs.

3 Role of iBALT in Immunity Against Infectious Diseases

The structure of iBALT suggests that it should promote primary immune responses against pulmonary antigens. In fact, antigen-specific T cell and B cell responses are initiated in iBALT, leading to B and T cell activation, germinal center formation, and the differentiation of plasma cells and effector T cells (Halle et al. 2009; Gregson et al. 1979; Shilling et al. 2013). The functional outcomes of these responses are often dependent on the type of pathogen or antigen as well as the quality of the resulting immune response. Below, we will summarize what we know about the role of iBALT in regulating immunity to different classes of pathogens.

a. **Mycobacterial and other bacterial infections**

The development or expansion of iBALT is often associated with bacterial infections (Baluk et al. 2014a; Khader et al. 2011; Jupelli et al. 2013; Chiavolini et al. 2010; Linge et al. 2017). For example, rats infected with *Pseudomonas aeruginosa* develop iBALT (Iwata and Sato 1991), as do pigs infected with *Salmonella oranienburg*, *Mycoplasma granularum*, or hemolytic *streptococcus* (Jericho et al. 1971a, b). In mice, pulmonary infection with *Pseudomonas aeruginosa* or *Staphylococcus aureus* promotes the development of iBALT, in part via the expression of CXCL12, CXCL13, and IL-17A (Frija-Masson et al. 2017), similar to that seen in other models.

A consistent feature of most bacterial infections is the recruitment of neutrophils, which likely enhance iBALT formation in a variety of ways. For example, neutrophils express cytokines like APRIL that activate B cells (Tecchio et al. 2014). Moreover, neutrophils secrete proteases and reactive oxygen species that trigger epithelial and mesenchymal cell activation (Meyer-Hoffert and Wiedow 2011). In fact, serine proteases made by neutrophils promote iBALT formation by causing damage and triggering the expression of inflammatory chemokines (Solleti et al. 2016). Activated neutrophils also produce neutrophil extracellular traps (NETs), which consist of granular components precipitated on ejected chromatin (Kaplan and Radic 2012). The NETs help trap and kill bacteria, but also cause damage and inflammation that facilitate iBALT formation (Sørensen and Borregaard 2016; Zhao et al. 2015).

iBALT formation is also associated with infection by *Mycobacterium tuberculosis*, the causative agent of pulmonary tuberculosis, which kills more than a million people per year worldwide and is rapidly acquiring antibiotic resistance (Orme et al. 2015). The course of disease is characterized by a temporary paralysis of DC migration to the lung-draining lymph nodes (Curtis et al. 2015; Vanessa et al. 2015; Lai et al. 2014; Roberts and Robinson 2014), which delays the generation of Th1 and Th17 responses (Doz et al. 2013; Demangel et al. 2002), thereby allowing the bacilli to accumulate in infected macrophages (Khan et al. 2019; Kang et al. 2011; Blomgran et al. 2012). Even when protective Th1 and Th17 responses are generated, *M. tuberculosis* survives, but is contained in a granuloma—a type of inducible lymphoid structure with a central area of infected macrophages surrounded by activated T cells and B cells (Cadena et al. 2017). These activated B and T cells often form iBALT surrounding the granulomas in *M. tuberculosis*-infected humans (Zhang et al. 2011; Ulrichs et al. 2004), non-human primates (Ganchua et al. 2018), and mice (Khader et al. 2011; Slight et al. 2013). Importantly, the presence of iBALT is associated with the maintenance of latency and containment of infection, whereas the absence of iBALT is associated with active disease (Ulrichs et al. 2004; Slight et al. 2013).

Although protective immunity against *M. tuberculosis* is mediated by IFNγ-producing Th1 cells, more recent data suggest IL-17A is also required (Khader et al. 2007, 2011; Doz et al. 2013; Martínez-Barricarte et al. 2018). IFN-γ activates macrophages and kills the bacilli, whereas IL-17A increases CXCL13 expression, which is required for the recruitment and organization of cellular infiltrates (Khader et al. 2007, 2011; Martínez-Barricarte et al. 2018; Gopal et al. 2013). Immune responses that deviate from these pathways fail to effectively control disease, as shown in mice previously exposed to *Schistosoma mansoni* egg antigen (SEA), which triggers a mixed Th1/Th2 response and thereby shifts the leukocyte infiltrate from B cell follicles to perivascular T cells and ultimately fails to control *M. tuberculosis* (DiNardo et al. 2016; Monin et al. 2015). Thus, effective immunity to *M. tuberculosis* requires the proper spatial positioning of cells in the lung consistent with iBALT formation.

Given the apparent protective effects of iBALT in the context of pulmonary infections, it makes sense to develop pulmonary vaccines that also trigger iBALT formation (Sanchez-Guzman et al. 2019). For example, pulmonary vaccination with *Francisella tularensis* LPS as a vaccine antigen and recombinant Porin B as an adjuvant promotes iBALT formation and germinal center development, leading to significant titers of LPS-reactive IgG and IgM that, together with iBALT, protect the immunized mice from subsequent challenge infection (Chiavolini et al. 2010). Similarly, the pulmonary administration of protein nanoparticles promotes iBALT formation in an antigen-non-specific fashion, leading to improved immune outcomes following pulmonary infection with the intracellular bacteria *Coxiella burnetii* (Wiley et al. 2009). Thus, the formation of iBALT in response to antigen-specific and antigen-non-specific stimuli provide subsequent protection from bacterial infections.

b. **Viral infections**

Pulmonary infection with viruses, including influenza (GeurtsvanKessel et al. 2009; Denton et al. 2019b; Richert et al. 2013), MVA (Fleige and Forster 2017; Fleige et al. 2018; Mzinza et al. 2018), respiratory syncytial virus (RSV) (Auais et al. 2003), SARS coronavirus (Channappanavar et al. 2014) and adenovirus (Jericho et al. 1971b), is often associated with the formation of iBALT. In mice, influenza infection promotes iBALT formation, which supports germinal center responses and the local differentiation of influenza-specific plasma cells (GeurtsvanKessel et al. 2009; Rangel-Moreno et al. 2011), many of which differentiate locally, as the disruption of iBALT two weeks after infection reduces local IgA production (GeurtsvanKessel et al. 2009). Moreover, influenza-specific memory B cells in the lung are more broadly reactive against numerous strains of influenza (Adachi et al. 2015), suggesting that the BCR selection process in the germinal centers of iBALT is qualitatively different than that in LNs. Moreover, mice with pre-existing iBALT experience an accelerated, influenza-specific antibody response in the lung (Rangel-Moreno et al. 2011; Wiley et al. 2009) and perform better than control mice in terms of weight loss and viral titers. Interestingly, iBALT also forms in the lungs of influenza-infected adult monkeys, but not in influenza-infected infants (Holbrook et al. 2015), leading to poor antibody responses and increased pulmonary damage in infants.

The presence of iBALT also provides a beneficial effect with SARS coronavirus, which is cleared more rapidly in mice with iBALT by an accelerated antibody response (Wiley et al. 2009). Similarly, mice that have iBALT induced as a result of neonatal LPS exposure lose less weight and clear pneumovirus faster than mice without iBALT (Foo et al. 2015). Importantly, $CD4^+$ T cell response to pneumovirus is accelerated in mice with iBALT (Foo et al. 2015), suggesting that the presence of iBALT in the lung leads to faster, more efficient pulmonary immune responses that promote rapid viral clearance and reduce morbidity after infection.

Although a faster more robust immune response may be desirable for immunity to many pathogens, some viruses elicit immune responses that are themselves the primary cause of pathogenesis. For example, RSV causes acute bronchiolitis in children and is linked to recurrent wheezing and asthma (Munywoki et al. 2013). Interestingly, infection of CCR7-deficient mice with RSV leads to enhanced production of IL-17 and IL-13 by $CD4^+$ T cells and excessive mucus production (Kallal et al. 2010). RSV-infected LTα-deficient mice, which lack conventional lymphoid organs, also experience excessive IL-17 and IL-13 expression and increased mucus production in the lung, suggesting that local immune responses in iBALT are responsible for pathology (Kallal et al. 2010). Similar exacerbations of pulmonary pathology are linked to the presence of iBALT in RSV-infected humans (Johnson et al. 2007). The combination of a pulmonary allergic response and RSV infection is particularly damaging in guinea pigs, which develop exacerbated iBALT hyperplasia, goblet cell metaplasia, and airway hypersensitivity (Robinson et al. 1997). Thus, in the context of RSV and perhaps other Th2-driven pulmonary conditions, the presence of iBALT may exacerbate disease simply by driving bigger, better faster immune responses that are more pathologic than protective.

c. **Fungal infections**

Mice infected with the opportunistic fungal pathogen, *Pneumocystis*, often generate a mixed Th17/Th2 response. Importantly, the combination of IL-13 and IL-17 synergistically promotes the differentiation of pulmonary fibroblasts and their expression of CXCL13, ultimately leading to iBALT formation (Eddens et al. 2017). Activated DCs also accumulate in the lungs of *Pneumocystis*-infected mice and potentiate T cell priming to other pulmonary antigens (Swain et al. 2011). In fact, prior infection with *Pneumocystis* enhances subsequent immunity to the influenza virus, leading to the accelerated appearance of influenza-specific antibodies and reduced expression of inflammatory cytokines in the bronchoalveolar lavage fluid, thereby reducing morbidity and accelerating viral clearance (Wiley and Harmsen 2008). Thus, the formation of iBALT in response to one pathogen enhances immunity to unrelated pathogens.

4 Role of iBALT in the Immune Response Against Non-infectious Agents

a. **Allergens**

Allergic or atopic immune responses are mediated by inappropriate Th2 and/or Th17 responses against non-pathogenic, environmental antigens, such as food antigens (peanut, egg), arthropods (house dust mite, cockroach), and plant components (pollen). The frequency of individuals developing severe allergies or asthma is rapidly increasing for unknown reasons (Jappe et al. 2019). Allergic responses typically involve a sensitization phase, in which allergen exposure primes T cells, but does not cause symptoms (Pizzolla et al. 2016; Shilovskiy et al. 2019), and a challenge phase, in which exposure to the same allergen caused an atopic inflammatory response (Shinoda et al. 2017; Gregory and Lloyd 2011). In the lung, chronic allergic responses promote airway remodeling, goblet cell hyperplasia and excessive mucus production, ultimately leading to reductions in lung function (Elieh Ali Komi and Bjermer 2019; Holt and Sly 2007) and obstructive leukocyte infiltration (Lainez et al. 2019; Maselli and Hanania 2019).

Chronic or repetitive exposure to allergens can trigger iBALT formation (Guest and Sell 2015). Hypersensitivity pneumonitis (sometimes called farmer's lung) is a classic example, in which repeated exposure to molds or other antigens in barn dust leads to lung disease, in which iBALT features prominently (Suda et al. 1999). The inflammatory milieu of allergic responses supports iBALT formation via numerous mechanisms, including the combined expression of IL-13 and IL-17 that promote stromal cell differentiation (Eddens et al. 2017). Moreover, Th2-related cytokines like IL-5 promote the recruitment of eosinophils, which likely accelerate iBALT formation by releasing granular contents including proteases and cytokines that in turn cause damage and support cellular differentiation (Lee et al. 1997a; b). In fact, this process can be mimicked by the overexpression of IL-5 in club cells (Lee et al. 1997a), which promotes eosinophil accumulation and iBAT formation in the absence of exogenous antigen.

The presence of iBALT in the lungs might contribute to the development of allergies by preferentially recruiting Th2 memory cells into the lung (Fleige et al. 2018; Shinoda et al. 2016), by increasing the concentration of IL-33 due to the differentiation of new lymphatic endothelial cells (Shinoda et al. 2016, 2017), or by supporting germinal centers that produce IgE^+ or $IgG1^+$ plasma cells (Chvatchko et al. 1996). In fact, iBALT may generally exacerbate atopic inflammation by supporting bigger, better, faster (albeit inappropriate) immune responses in the lung. One way to accomplish this goal would be to recruit $Gata3^+CXCR5^+$ T follicular helper (Tfh13) cells to iBALT (Noble and Zhao 2016). Tfh13 cells strongly produce IL-4, IL-5, and IL-13, but not IL-21, conditions that support B cell differentiation into antibody-secreting cells that make IgG1 or high-affinity IgE (Gowthaman et al. 2019).

Although repeated allergen exposure can lead to eosinophil recruitment, mucus production, and IgE secretion, thereby promoting allergic inflammation and pathology, these same activities should help control parasitic infection. In fact, mice pre-sensitized with house dust mite extract developed iBALT areas, recruited eosinophils, and expressed high levels of IL-4, IL-13, and IL-33 in the lungs, which together acted to prevent the maturation of *Ascaris* larvae, whereas mice not pre-sensitized with house dust mite failed to prevent larval development (Gazzinelli-Guimaraes et al. 2019). These data suggest that although Th2-driven iBALT formation may enhance atopic responses and promote pulmonary inflammation, it may also be beneficial in the clearance of pulmonary parasites.

b. **Self-antigens: the good and the bad, can we tell them apart?**

TLOs, including iBALT, are often formed around tumors, in transplanted organs and in the target organs of autoimmune responses. For example, the presence of iBALT near tumor nests in patients with non-small-cell lung cancer (NSCLC) correlates with a better prognosis (Dieu-Nosjean et al. 2016). Within iBALT, higher numbers of DCs in close proximity to tumor cells (Dieu-Nosjean et al. 2008), the presence of $Tbet^{+}CD4^{+}$ T cells (Goc et al. 2014), and the frequency of $CD161^{+}CD4^{+}$ T cells (Braud et al. 2018), all indicate an active anti-tumor response and correlate with better clinical outcomes. For some tumors, including breast cancer (Peske et al. 2015), ovarian cancer (Kroeger et al. 2016; Truxova et al. 2018), and NSCLC (Germain et al. 2014), the presence of TLOs is associated with a favorable prognosis, whereas in tumors like colorectal cancer, the chronic inflammation associated with TLO formation is also linked to tumorigenesis (Weinstein et al. 2019).

TLO development is initially triggered inflammatory responses that promote the activation and differentiation of mesenchymal cells and trigger the expression of CXCL13. In tumor that lack microbial components, inflammatory signals might come from the release of danger-associated molecular patterns (DAMPs), such as IL-1α (Kuroda et al. 2016), IL-18 (Briend et al. 2017), or IL-36γ (Weinstein et al. 2017). Increased CXCL13 expression and the recruitment of LT-expressing lymphocytes promote the expression of ICAM, VCAM, PNAd, and CCL21 in blood endothelial cells (BEC) and reinforce the recruitment of more LT-bearing cells. The continuous signaling of LT$\alpha\beta$-LTβR activates the non-canonical NF-kB pathway and leads to the differentiation of BEC into $PNAd^{+}$ HEV (Ager 2017). Can the process of HEV development be exploited to improve immune responses against tumor cells? In this regard, VEGFR2 blockade prevents angiogenesis in tumors, but also induces PD-L1 expression by tumor cells, thus impairing anti-tumor immunity. However, the combined blockade of VEGFR2 and PD-L1 antibody maintained anti-tumor immunity and promoted the differentiation of BEC into HEVs by the constant influx of activated LT-expressing lymphocytes (Allen et al. 2017). Thus, the mechanisms that regulate lymphocyte recruitment and TLO formation can be exploited for therapeutic benefit.

In contrast to the beneficial effect of local immunity against tumors, local immune responses against organ transplants, including transplanted lungs, can lead to graft rejection (Kumar et al. 2018). Not surprisingly the formation of iBALT with active germinal centers is indicative of an ongoing immune response against transplanted lungs and is associated with the development of antibody-mediated rejection (Gauthier et al. 2019; Shenoy et al. 2012; Hasegawa et al. 1999). Interestingly, this process can be prevented by the recruitment of Tregs, which suppress germinal center formation in iBALT and prevent allo-antibody production (Li et al. 2019). The switch from immunity to tolerance is mediated by the blockade of costimulatory signals through CD40 and CD28. Moreover, once Tregs are recruited to iBALT areas in the transplanted lung, it can be re-transplanted to another recipient without rejection (Li et al. 2019)! Importantly, $CXCR5^+$ Tregs in limiting lung rejection after chronic GVHD were demonstrated in B10.BR mice receiving lungs from C57BL/6 donors (McDonald-Hyman et al. 2016). In the recipient B10.BR mice, lung transplants improved their function and reduced the number of T follicular helper cells when receiving a passive transfer of $CXCR5^+$ Tregs but not with $CXCR5^-$ Tregs. Overall these studies suggest that iBALT facilitates the entry and interaction of $CXCR5^+$ lymphocytes and that the type of local immune response depends on the lymphocyte subsets recruited (Li et al. 2019; McDonald-Hyman et al. 2016; Flynn et al. 2014).

Furthermore, in autoimmune diseases like rheumatoid arthritis (RA) and Wegener's granulomatosis (WG), iBALT can develop in the lungs and its occurrence is associated with a chronic and worsening status of the disease (Shilling et al. 2013). For instance, in RA increased concentration of serum rheumatoid factor (IgM antibodies directed against IgG Fc portion) correlates with the appearance of rheumatoid pulmonary vasculitis and TLO in the lungs (Rangel-Moreno et al. 2006). In Wegener's granulomatosis (WG), lymphocytes in the lungs can form diffuse infiltrates, but also can form structured iBALT and form germinal centers (Shilling et al. 2013). The pronounced infiltration of granulocytes is characteristic of a Th17-driven disease and is consistent with the role of IL-17 in promoting iBALT formation.

5 Conclusion

Like many tertiary lymphoid organs, iBALT forms in response to a variety of inflammatory stimuli that converge on the differentiation of specialized stromal cells, the expression of homeostatic chemokines and the recruitment and organization of activated lymphocytes. Once formed, iBALT participates in local, pulmonary immune responses by collecting antigen and APCs and supporting B and T cell responses. The biological outcome of those immune responses is on the type of antigen or pathogen and may be modified by the presence of iBALT by changing the kinetics or magnitude of the resulting immune response, which may be beneficial or harmful depending on the context. Thus, understanding the mechanisms that control iBALT formation and function should give us insights into ways to improve immunity to pathogens and malignancy and to dampen atopic or inflammatory diseases.

References

Adachi Y, Onodera T, Yamada Y, Daio R, Tsuiji M, Inoue T, Kobayashi K, Kurosaki T, Ato M, Takahashi Y (2015) Distinct germinal center selection at local sites shapes memory B cell response to viral escape. J Exp Med 212:1709–1723

Ager A (2017) High endothelial venules and other blood vessels: critical regulators of lymphoid organ development and function. Front Immunol 8:45

Allen E, Jabouille A, Rivera LB, Lodewijckx I, Missiaen R, Steri V, Feyen K, Tawney J, Hanahan D, Michael IP, Bergers G (2017) Combined antiangiogenic and anti-PD-L1 therapy stimulates tumor immunity through HEV formation. Sci Transl Med 9

Astarita JL, Cremasco V, Fu J, Darnell MC, Peck JR, Nieves-Bonilla JM, Song K, Kondo Y, Woodruff MC, Gogineni A, Onder L, Ludewig B, Weimer RM, Carroll MC, Mooney DJ, Xia L, Turley SJ (2015) The CLEC-2-podoplanin axis controls the contractility of fibroblastic reticular cells and lymph node microarchitecture. Nat Immunol 16:75–84

Auais A, Adkins B, Napchan G, Piedimonte G (2003) Immunomodulatory effects of sensory nerves during respiratory syncytial virus infection in rats. Am J Physiol Lung Cell Mol Physiol 285:L105–L113

Aujla SJ, Kolls JK (2009) IL-22: a critical mediator in mucosal host defense. J Mol Med (Berl) 87:451–454

Aziz KE, McCluskey PJ, Wakefield D (1997) Characterisation of follicular dendritic cells in labial salivary glands of patients with primary Sjögren syndrome: comparison with tonsillar lymphoid follicles. Ann Rheum Dis 56:140–143

Baluk P, Yao LC, Feng J, Romano T, Jung SS, Schreiter JL, Yan L, Shealy DJ, McDonald DM (2009) TNF-alpha drives remodeling of blood vessels and lymphatics in sustained airway inflammation in mice. J Clin Invest 119:2954–2964

Baluk P, Hogmalm A, Bry M, Alitalo K, Bry K, McDonald DM (2013) Transgenic overexpression of interleukin-1β induces persistent lymphangiogenesis but not angiogenesis in mouse airways. Am J Pathol 182:1434–1447

Baluk P, Adams A, Phillips K, Feng J, Hong YK, Brown MB, McDonald DM (2014a) Preferential lymphatic growth in bronchus-associated lymphoid tissue in sustained lung inflammation. Am J Pathol 184:1577–1592

Baluk P, Phillips K, Yao LC, Adams A, Nitschké M, McDonald DM (2014b) Neutrophil dependence of vascular remodeling after mycoplasma infection of mouse airways. Am J Pathol 184:1877–1889

Baratin M, Foray C, Demaria O, Habbeddine M, Pollet E, Maurizio J, Verthuy C, Davanture S, Azukizawa H, Flores-Langarica A, Dalod M, Lawrence T (2015) Homeostatic NF-kappaB signaling in steady-state migratory dendritic cells regulates immune homeostasis and tolerance. Immunity 42:627–639

Barone F, Nayar S, Campos J, Cloake T, Withers DR, Toellner KM, Zhang Y, Fouser LFisher B, Bowman S, Rangel-Moreno J, Garcia-Hernandez ML, Randall TD, Lucchesi D, Bombardieri M, Pitzalis C, Luther SA, Buckley CD (2015) IL-22 regulates lymphoid chemokine production and assembly of tertiary lymphoid organs. Proc Natl Acad Sci USA 112:11024–11029

Berrih-Aknin S, Ragheb S, Le Panse R, Lisak RP (2013) Ectopic germinal centers, BAFF and anti-B-cell therapy in myasthenia gravis. Autoimmun Rev 12:885–893

Blomgran R, Desvignes L, Briken V, Ernst JD (2012) *Mycobacterium tuberculosis* inhibits neutrophil apoptosis, leading to delayed activation of naive CD4 T cells. Cell Host Microbe 11:81–90

Botelho FM, Rangel-Moreno J, Fritz D, Randall TD, Xing Z, Richards CD (2013) Pulmonary expression of oncostatin M (OSM) promotes inducible BALT formation independently of IL-6, despite a role for IL-6 in OSM-driven pulmonary inflammation. J Immunol 191:1453–1464

Bouton MC, Boulaftali Y, Richard B, Arocas V, Michel JB, Jandrot-Perrus M (2012) Emerging role of serpinE2/protease nexin-1 in hemostasis and vascular biology. Blood 119:2452–2457

Braud VM, Biton J, Becht E, Knockaert S, Mansuet-Lupo A, Cosson E, Damotte D, Alifano M, Validire P, Anjuère F, Cremer I, Girard N, Gossot D, Seguin-Givelet A, Dieu-Nosjean MC, Germain C (2018) Expression of LLT1 and its receptor CD161 in lung cancer is associated with better clinical outcome. Oncoimmunology 7:e1423184

Briend E, Ferguson GJ, Mori M, Damera G, Stephenson K, Karp NA, Sethi S, Ward CK, Sleeman MA, Erjefält JS, Finch DK (2017) IL-18 associated with lung lymphoid aggregates drives IFNγ production in severe COPD. Respir Res 18:159

Browning JL, Allaire N, Ngam-Ek A, Notidis E, Hunt J, Perrin S, Fava RA (2005) Lymphotoxin-beta receptor signaling is required for the homeostatic control of HEV differentiation and function. Immunity 23:539–550

Cadena AM, Fortune SM, Flynn JL (2017) Heterogeneity in tuberculosis. Nat Rev Immunol 17:691–702

Carlsen HS, Baekkevold ES, Johansen FE, Haraldsen G, Brandtzaeg P (2002) B cell attracting chemokine 1 (CXCL13) and its receptor CXCR5 are expressed in normal and aberrant gut associated lymphoid tissue. Gut 51:364–371

Channappanavar R, Fett C, Zhao J, Meyerholz DK, Perlman S (2014) Virus-specific memory CD8 T cells provide substantial protection from lethal severe acute respiratory syndrome coronavirus infection. J Virol 88:11034–11044

Chiavolini D, Rangel-Moreno J, Berg G, Christian K, Oliveira-Nascimento L, Weir S, Alroy J, Randall TD, Wetzler LM (2010) Bronchus-associated lymphoid tissue (BALT) and survival in a vaccine mouse model of tularemia. PLoS ONE 5:e11156

Chvatchko Y, Kosco-Vilbois MH, Herren S, Lefort J, Bonnefoy JY (1996) Germinal center formation and local immunoglobulin E (IgE) production in the lung after an airway antigenic challenge. J Exp Med 184:2353–2360

Ciccia F, Rizzo A, Maugeri R, Alessandro R, Croci S, Guggino G, Cavazza A, Raimondo S, Cannizzaro A, Iacopino DG, Salvarani C, Triolo G (2017) Ectopic expression of CXCL13, BAFF, APRIL and LT-β is associated with artery tertiary lymphoid organs in giant cell arteritis. Ann Rheum Dis 76:235–243

Cowan JE, Parnell SM, Nakamura K, Caamano JH, Lane PJ, Jenkinson EJ, Jenkinson WE, Anderson G (2013) The thymic medulla is required for Foxp3+ regulatory but not conventional CD4+ thymocyte development. J Exp Med 210:675–681

Cupedo T, Jansen W, Kraal G, Mebius RE (2004) Induction of secondary and tertiary lymphoid structures in the skin. Immunity 21:655–667

Curtis J, Luo Y, Zenner HL, Cuchet-Lourenço D, Wu C, Lo K, Maes M, Alisaac A, Stebbings E, Liu JZ, Kopanitsa L, Ignatyeva O, Balabanova Y, Nikolayevskyy V, Baessmann I, Thye T, Meyer CG, Nürnberg P, Horstmann RD, Drobniewski F, Plagnol V, Barrett JC, Nejentsev S (2015) Susceptibility to tuberculosis is associated with variants in the ASAP1 gene encoding a regulator of dendritic cell migration. Nat Genet 47:523–527

Das A, Kole L, Wang L, Barrios R, Moorthy B, Jaiswal AK (2006) BALT development and augmentation of hyperoxic lung injury in mice deficient in NQO1 and NQO2. Free Radic Biol Med 40:1843–1856

Demangel C, Bertolino P, Britton WJ (2002) Autocrine IL-10 impairs dendritic cell (DC)-derived immune responses to mycobacterial infection by suppressing DC trafficking to draining lymph nodes and local IL-12 production. Eur J Immunol 32:994–1002

Denton AE, Carr EJ, Magiera LP, Watts AJB, Fearon DT (2019) Embryonic FAP. J Exp Med
Denton AE, Innocentin S, Carr EJ, Bradford BM, Lafouresse F, Mabbott NA, Mörbe U, Ludewig B, Groom JR, Good-Jacobson KL, Linterman MA (2019b) Type I interferon induces CXCL13 to support ectopic germinal center formation. J Exp Med 216:621–637
Dieu-Nosjean MC, Antoine M, Danel C, Heudes D, Wislez M, Poulot V, Rabbe N, Laurans L, Tartour E, de Chaisemartin L, Lebecque S, Fridman WH, Cadranel J (2008) Long-term survival for patients with non-small-cell lung cancer with intratumoral lymphoid structures. J Clin Oncol 26:4410–4417
Dieu-Nosjean MC, Giraldo NA, Kaplon H, Germain C, Fridman WH, Sautès-Fridman C (2016) Tertiary lymphoid structures, drivers of the anti-tumor responses in human cancers. Immunol Rev 271:260–275
DiNardo AR, Mace EM, Lesteberg K, Cirillo JD, Mandalakas AM, Graviss EA, Orange JS, Makedonas G (2016) Schistosome soluble egg antigen decreases *Mycobacterium tuberculosis*-specific CD4+ T-cell effector function with concomitant arrest of macrophage phago-lysosome maturation. J Infect Dis 214:479–488
Ding J, Zhang W, Haskett S, Pellerin A, Xu S, Petersen B, Jandreski L, Hamann S, Reynolds TL, Zheng TS, Mingueneau M (2016) BAFF overexpression increases lymphocytic infiltration in Sjögren's target tissue, but only inefficiently promotes ectopic B-cell differentiation. Clin Immunol 169:69–79
Doz E, Lombard R, Carreras F, Buzoni-Gatel D, Winter N (2013) Mycobacteria-infected dendritic cells attract neutrophils that produce IL-10 and specifically shut down Th17 CD4 T cells through their IL-10 receptor. J Immunol 191:3818–3826
Drayton DL, Ying X, Lee J, Lesslauer W, Ruddle NH (2003) Ectopic LT alpha beta directs lymphoid organ neogenesis with concomitant expression of peripheral node addressin and a HEV-restricted sulfotransferase. J Exp Med 197:1153–1163
Eddens T, Elsegeiny W, Garcia-Hernadez ML, Castillo P, Trevejo-Nunez G, Serody K, Campfield BT, Khader SA, Chen K, Rangel-Moreno J, Kolls JK (2017) Pneumocystis-driven inducible bronchus-associated lymphoid tissue formation requires Th2 and Th17 immunity. Cell Rep 18:3078–3090
Elieh Ali Komi, D. and Bjermer, L. (2019) Mast cell-mediated orchestration of the immune responses in human allergic asthma: current insights. Clin Rev Allergy Immunol 56:234–247
Flajnik MF (2002) Comparative analyses of immunoglobulin genes: surprises and portents. Nat Rev Immunol 2:688–698
Fleige H, Forster R (2017) Induction and analysis of bronchus-associated lymphoid tissue. Methods Mol Biol 1559:185–198
Fleige H, Haas JD, Stahl FR, Willenzon S, Prinz I, Forster R (2012) Induction of BALT in the absence of IL-17. Nat Immunol 13, 1; author reply 2
Fleige H, Ravens S, Moschovakis GL, Bolter J, Willenzon S, Sutter G, Haussler S, Kalinke U, Prinz I, Forster R (2014) IL-17-induced CXCL12 recruits B cells and induces follicle formation in BALT in the absence of differentiated FDCs. J Exp Med 211:643–651
Fleige H, Bosnjak B, Permanyer M, Ristenpart J, Bubke A, Willenzon S, Sutter G, Luther SA, Förster R (2018) Manifold roles of CCR7 and its ligands in the induction and maintenance of bronchus-associated lymphoid tissue. Cell Rep 23:783–795
Fletcher AL, Malhotra D, Acton SE, Lukacs-Kornek V, Bellemare-Pelletier A, Curry M, Armant M, Turley SJ (2011) Reproducible isolation of lymph node stromal cells reveals site-dependent differences in fibroblastic reticular cells. Front Immunol 2:35
Flynn R, Du J, Veenstra RG, Reichenbach DK, Panoskaltsis-Mortari A, Taylor PA, Freeman GJ, Serody JS, Murphy WJ, Munn DH, Sarantopoulos S, Luznik L, Maillard I, Koreth J, Cutler C, Soiffer RJ, Antin JH, Ritz J, Dubovsky JA, Byrd JC, MacDonald KP, Hill GR, Blazar BR (2014) Increased T follicular helper cells and germinal center B cells are required for cGVHD and bronchiolitis obliterans. Blood 123:3988–3998
Foo SY, Zhang V, Lalwani A, Lynch JP, Zhuang A, Lam CE, Foster PS, King C, Steptoe RJ, Mazzone SB, Sly PD, Phipps S (2015) Regulatory T cells prevent inducible BALT formation by dampening neutrophilic inflammation. J Immunol 194:4567–4576

Frija-Masson J, Martin C, Regard L, Lothe MN, Touqui L, Durand A, Lucas B, Damotte D, Alifano M, Fajac I, Burgel PR (2017) Bacteria-driven peribronchial lymphoid neogenesis in bronchiectasis and cystic fibrosis. Eur Respir J 49

Furtado GC, Pacer ME, Bongers G, Benezech C, He Z, Chen L, Berin MC, Kollias G, Caamano JH, Lira SA (2014) TNFalpha-dependent development of lymphoid tissue in the absence of RORgammat(+) lymphoid tissue inducer cells. Mucosal Immunol 7:602–614

Ganchua SKC, Cadena AM, Maiello P, Gideon HP, Myers AJ, Junecko BF, Klein EC, Lin PL, Mattila JT, Flynn JL (2018) Lymph nodes are sites of prolonged bacterial persistence during Mycobacterium tuberculosis infection in macaques. PLoS Pathog 14:e1007337

Gauthier JM, Harrison MS, Krupnick AS, Gelman AE, Kreisel D (2019) The emerging role of regulatory T cells following lung transplantation. Immunol Rev

Gazzinelli-Guimaraes PH, de Queiroz Prado R, Ricciardi A, Bonne-Année S, Sciurba J, Karmele EP, Fujiwara RT, Nutman TB (2019) Allergen presensitization drives an eosinophil-dependent arrest in lung-specific helminth development. J Clin Invest 130:3686–3701

Gentek R, Bajenoff M (2017) Lymph node stroma dynamics and approaches for their visualization. Trends Immunol 38:236–247

Gentilini MV, Nunez GG, Roux ME, Venturiello SM (2011) Trichinella spiralis infection rapidly induces lung inflammatory response: the lung as the site of helminthocytotoxic activity. Immunobiology 216:1054–1063

Georgiev P, Charbonnier LM, Chatila TA (2019) Regulatory T cells: the many faces of Foxp3. J Clin Immunol 39:623–640

Germain C, Gnjatic S, Tamzalit F, Knockaert S, Remark R, Goc J, Lepelley A, Becht E, Katsahian S, Bizouard G, Validire P, Damotte D, Alifano M, Magdeleinat P, Cremer I, Teillaud JL, Fridman WH, Sautès-Fridman C, Dieu-Nosjean MC (2014) Presence of B cells in tertiary lymphoid structures is associated with a protective immunity in patients with lung cancer. Am J Respir Crit Care Med 189:832–844

GeurtsvanKessel CH, Willart MA, Bergen IM, van Rijt LS, Muskens F, Elewaut D, Osterhaus AD, Hendriks R, Rimmelzwaan GF, Lambrecht BN (2009) Dendritic cells are crucial for maintenance of tertiary lymphoid structures in the lung of influenza virus-infected mice. J Exp Med 206:2339–2349

Girard JP, Moussion C, Forster R (2012) HEVs, lymphatics and homeostatic immune cell trafficking in lymph nodes. Nat Rev Immunol 12:762–773

Goc J, Germain C, Vo-Bourgais TK, Lupo A, Klein C, Knockaert S, de Chaisemartin L, Ouakrim H, Becht E, Alifano M, Validire P, Remark R, Hammond SA, Cremer I, Damotte D, Fridman WH, Sautès-Fridman C, Dieu-Nosjean MC (2014) Dendritic cells in tumor-associated tertiary lymphoid structures signal a Th1 cytotoxic immune contexture and license the positive prognostic value of infiltrating CD8+ T cells. Cancer Res 74:705–715

Gopal R, Rangel-Moreno J, Slight S, Lin Y, Nawar HF, Fallert Junecko BA, Reinhart TA, Kolls J, Randall TD, Connell TD, Khader SA (2013) Interleukin-17-dependent CXCL13 mediates mucosal vaccine-induced immunity against tuberculosis. Mucosal Immunol 6:972–984

Gowthaman U, Chen JS., Zhang B, Flynn WF, Lu Y, Song W, Joseph J, Gertie JA, Xu L, Collet MA, Grassmann JDS, Simoneau T, Chiang D, Berin MC, Craft JE, Weinstein JS, Williams A, Eisenbarth SC (2019) Identification of a T follicular helper cell subset that drives anaphylactic IgE. Science 365

Gregory LG, Lloyd CM (2011) Orchestrating house dust mite-associated allergy in the lung. Trends Immunol 32:402–411

Gregson RL, Davey MJ, Prentice DE (1979) The response of rat bronchus-associated lymphoid tissue to local antigenic challenge. Br J Exp Pathol 60:471–482

Guedj K, Khallou-Laschet J, Clement M, Morvan M, Gaston AT, Fornasa G, Dai J, Gervais-Taurel M, Eberl G, Michel JB, Caligiuri G, Nicoletti A (2014) M1 macrophages act as LTbetaR-independent lymphoid tissue inducer cells during atherosclerosis-related lymphoid neogenesis. Cardiovasc Res 101:434–443

Guest IC, Sell S (2015) Bronchial lesions of mouse model of asthma are preceded by immune complex vasculitis and induced bronchial associated lymphoid tissue (iBALT). Lab Invest 95:886–902

Halle S, Dujardin HC, Bakocevic N, Fleige H, Danzer H, Willenzon S, Suezer Y, Hammerling G, Garbi N, Sutter G, Worbs T, Forster R (2009) Induced bronchus-associated lymphoid tissue serves as a general priming site for T cells and is maintained by dendritic cells. J Exp Med 206:2593–2601

Hasegawa T, Iacono A, Yousem SA (1999) The significance of bronchus-associated lymphoid tissue in human lung transplantation: is there an association with acute and chronic rejection? Transplantation 67:381–385

Herzog BH, Fu J, Wilson SJ, Hess PR, Sen A, McDaniel JM, Pan Y, Sheng M, Yago T, Silasi-Mansat R, McGee S, May F, Nieswandt B, Morris AJ, Lupu F, Coughlin SR, McEver RP, Chen H, Kahn ML, Xia L (2013) Podoplanin maintains high endothelial venule integrity by interacting with platelet CLEC-2. Nature 502:105–109

Holbrook BC, Hayward SL, Blevins LK, Kock N, Aycock T, Parks GD, Alexander-Miller MA (2015) Nonhuman primate infants have an impaired respiratory but not systemic IgG antibody response following influenza virus infection. Virology 476:124–133

Holt PG (1993) Development of bronchus associated lymphoid tissue (BALT) in human lung disease: a normal host defence mechanism awaiting therapeutic exploitation? Thorax 48:1097–1098

Holt PG, Sly PD (2007) Th2 cytokines in the asthma late-phase response. The Lancet 370:1396–1398

Hwang JY, Randall TD, Silva-Sanchez A (2016) Inducible bronchus-associated lymphoid tissue: taming inflammation in the lung. Front Immunol 7:258

Iwata M, Sato A (1991) Morphological and immunohistochemical studies of the lungs and bronchus-associated lymphoid tissue in a rat model of chronic pulmonary infection with *Pseudomonas aeruginosa*. Infect Immun 59:1514–1520

Jappe U, Schwager C, Schromm AB, González Roldán N, Stein K, Heine H, Duda KA (2019) Lipophilic allergens, different modes of allergen-lipid interaction and their impact on asthma and allergy. Front Immunol 10:122

Jericho KW, Austwick PK, Hodges RT, Dixon JB (1971a) Intrapulmonary lymphoid tissue of pigs exposed to aerosols of carbon particles, of *Salmonella oranienburg*, of *Mycoplasma granularum*, and to an oral inoculum of larvae of Metastrongylus apri. J Comp Pathol 81:13–21

Jericho KW, Derbyshire JB, Jones JE (1971b) Intrapulmonary lymphoid tissue of pigs exposed to aerosols of haemolytic streptococcus group L and porcine adenovirus. J Comp Pathol 81:1–11

Johnson JE, Gonzales RA, Olson SJ, Wright PF, Graham BS (2007) The histopathology of fatal untreated human respiratory syncytial virus infection. Mod Pathol 20:108–119

Jupelli M, Shimada K, Chiba N, Slepenkin A, Alsabeh R, Jones HD, Peterson E, Chen S, Arditi M, Crother TR (2013) *Chlamydia pneumoniae* infection in mice induces chronic lung inflammation, iBALT formation, and fibrosis. PLoS ONE 8:e77447

Kallal LE, Hartigan AJ, Hogaboam CM, Schaller MA, Lukacs NW (2010) Inefficient lymph node sensitization during respiratory viral infection promotes IL-17-mediated lung pathology. J Immunol 185:4137–4147

Kang DD, Lin Y, Moreno JR, Randall TD, Khader SA (2011) Profiling early lung immune responses in the mouse model of tuberculosis. PLoS ONE 6:e16161

Kaplan MJ, Radic M (2012) Neutrophil extracellular traps: double-edged swords of innate immunity. J Immunol 189:2689–2695

Khader SA, Bell GK, Pearl JE, Fountain JJ, Rangel-Moreno J, Cilley GE, Shen F, Eaton SM, Gaffen SL, Swain SL, Locksley RM, Haynes L, Randall TD, Cooper AM (2007) IL-23 and IL-17 in the establishment of protective pulmonary CD4+ T cell responses after vaccination and during *Mycobacterium tuberculosis* challenge. Nat Immunol 8:369–377

Khader SA, Guglani L, Rangel-Moreno J, Gopal R, Junecko BA, Fountain JJ, Martino C, Pearl JE, Tighe M, Lin YY, Slight S, Kolls JK, Reinhart TA, Randall TD, Cooper AM (2011) IL-23 is required for long-term control of *Mycobacterium tuberculosis* and B cell follicle formation in the infected lung. J Immunol 187:5402–5407

Khan A, Singh VK, Hunter RL, Jagannath C (2019) Macrophage heterogeneity and plasticity in tuberculosis. J Leukoc Biol 106:275–282

Kocks JR, Davalos-Misslitz AC, Hintzen G, Ohl L, Forster R (2007) Regulatory T cells interfere with the development of bronchus-associated lymphoid tissue. J Exp Med 204:723–734

Kroeger DR, Milne K, Nelson BH (2016) Tumor-infiltrating plasma cells are associated with tertiary lymphoid structures, cytolytic T-cell responses, and superior prognosis in ovarian cancer. Clin Cancer Res 22:3005–3015

Kumar S, Leigh ND, Cao X (2018) The role of co-stimulatory/co-inhibitory signals in graft-vs.-host disease. Front Immunol 9:3003

Kuroda E, Ozasa K, Temizoz B, Ohata K, Koo CX, Kanuma T, Kusakabe T, Kobari S, Horie M, Morimoto Y, Nakajima S, Kabashima K, Ziegler SF, Iwakura Y, Ise W, Kurosaki T, Nagatake T, Kunisawa J, Takemura N, Uematsu S, Hayashi M, Aoshi T, Kobiyama K, Coban C, Ishii KJ (2016) Inhaled fine particles induce alveolar macrophage death and interleukin-1α release to promote inducible bronchus-associated lymphoid tissue formation. Immunity 45:1299–1310

Lai R, Jeyanathan M, Shaler CR, Damjanovic D, Khera A, Horvath C, Ashkar AA, Xing Z (2014) Restoration of innate immune activation accelerates Th1-cell priming and protection following pulmonary mycobacterial infection. Eur J Immunol 44:1375–1386

Lainez S, Court-Fortune I, Vercherin P, Falchero L, Didi T, Beynel P, Piperno D, Frappe E, Froudarakis M, Vergnon JM, Devouassoux G (2019) Clinical ACO phenotypes: description of a heterogeneous entity. Respir Med Case Rep 28:100929

Lee JJ, McGarry MP, Farmer SC, Denzler KL, Larson KA, Carrigan PE, Brenneise IE, Horton MA, Haczku A, Gelfand EW, Leikauf GD, Lee NA (1997a) Interleukin-5 expression in the lung epithelium of transgenic mice leads to pulmonary changes pathognomonic of asthma. J Exp Med 185:2143–2156

Lee NA, McGarry MP, Larson KA, Horton MA, Kristensen AB, Lee JJ (1997b) Expression of IL-5 in thymocytes/T cells leads to the development of a massive eosinophilia, extramedullary eosinophilopoiesis, and unique histopathologies. J Immunol 158:1332–1344

Li W, Gauthier JM, Higashikubo R, Hsiao HM, Tanaka S, Vuong L, Ritter JH, Tong AY, Wong BW, Hachem RR, Puri V, Bharat A, Krupnick AS, Hsieh CS, Baldwin WM, Kelly FL, Palmer SM, Gelman AE, Kreisel D (2019) Bronchus-associated lymphoid tissue-resident Foxp3+ T lymphocytes prevent antibody-mediated lung rejection. J Clin Invest 129:556–568

Linge I, Dyatlov A, Kondratieva E, Avdienko V, Apt A, Kondratieva T (2017) B-lymphocytes forming follicle-like structures in the lung tissue of tuberculosis-infected mice: dynamics, phenotypes and functional activity. Tuberculosis (Edinb) 102:16–23

Lu TT, Browning JL (2014) Role of the lymphotoxin/LIGHT system in the development and maintenance of reticular networks and vasculature in lymphoid tissues. Front Immunol 5:47

Luther SA, Ansel KM, Cyster JG (2003) Overlapping roles of CXCL13, interleukin 7 receptor alpha, and CCR7 ligands in lymph node development. J Exp Med 197:1191–1198

Marinkovic T, Garin A, Yokota Y, Fu YX, Ruddle NH, Furtado GC, Lira SA (2006) Interaction of mature CD3+ CD4+ T cells with dendritic cells triggers the development of tertiary lymphoid structures in the thyroid. J Clin Invest 116:2622–2632

Martinet L, Filleron T, Le Guellec S, Rochaix P, Garrido I, Girard JP (2013) High endothelial venule blood vessels for tumor-infiltrating lymphocytes are associated with lymphotoxin β-producing dendritic cells in human breast cancer. J Immunol 191:2001–2008

Martínez-Barricarte R, Markle JG, Ma CS, Deenick EK, Ramírez-Alejo N, Mele F, Latorre D, Mahdaviani SA, Aytekin C, Mansouri D, Bryant VL, Jabot-Hanin F, Deswarte C, Nieto-Patlán A, Surace L, Kerner G, Itan Y, Jovic S, Avery DT, Wong N, Rao G, Patin E, Okada S, Bigio B, Boisson B, Rapaport F, Seeleuthner Y, Schmidt M, Ikinciogullari A, Dogu F, Tanir G, Tabarsi P, Bloursaz MR, Joseph JK, Heer A, Kong XF, Migaud M, Lazarov T, Geissmann F,

Fleckenstein B, Arlehamn CL, Sette A, Puel A, Emile JF, van de Vosse E, Quintana-Murci L, Di Santo JP, Abel L, Boisson-Dupuis S, Bustamante J, Tangye SG, Sallusto F, Casanova JL (2018) Human IFN-γ immunity to mycobacteria is governed by both IL-12 and IL-23. Sci Immunol 3

Maselli DJ, Hanania NA (2019) Management of asthma COPD overlap. Ann Allergy Asthma Immunol

McDonald-Hyman C, Flynn R, Panoskaltsis-Mortari A, Peterson N, MacDonald KP, Hill GR, Luznik L, Serody JS, Murphy WJ, Maillard I, Munn DH, Turka LA, Koreth J, Cutler CS, Soiffer RJ, Antin JH, Ritz J, Blazar BR (2016) Therapeutic regulatory T-cell adoptive transfer ameliorates established murine chronic GVHD in a CXCR5-dependent manner. Blood 128:1013–1017

Mebius RE, Bauer J, Twisk AJ, Brevé J, Kraal G (1991a) The functional activity of high endothelial venules: a role for the subcapsular sinus macrophages in the lymph node. Immunobiology 182:277–291

Mebius RE, Streeter PR, Brevé J, Duijvestijn AM, Kraal G (1991b) The influence of afferent lymphatic vessel interruption on vascular addressin expression. J Cell Biol 115:85–95

Meyer-Hoffert U, Wiedow O (2011) Neutrophil serine proteases: mediators of innate immune responses. Curr Opin Hematol 18:19–24

Monin L, Griffiths KL, Lam WY, Gopal R, Kang DD, Ahmed M, Rajamanickam A, Cruz-Lagunas A, Zúñiga J, Babu S, Kolls JK, Mitreva M, Rosa BA, Ramos-Payan R, Morrison TE, Murray PJ, Rangel-Moreno J, Pearce EJ, Khader SA (2015) Helminth-induced arginase-1 exacerbates lung inflammation and disease severity in tuberculosis. J Clin Invest 125:4699–4713

Mounzer RH, Svendsen OS, Baluk P, Bergman CM, Padera TP, Wiig H, Jain RK, McDonald DM, Ruddle NH (2010) Lymphotoxin-alpha contributes to lymphangiogenesis. Blood 116:2173–2182

Moussion C, Girard JP (2011) Dendritic cells control lymphocyte entry to lymph nodes through high endothelial venules. Nature 479:542–546

Moyron-Quiroz JE, Rangel-Moreno J, Kusser K, Hartson L, Sprague F, Goodrich S, Woodland DL, Lund FE, Randall TD (2004) Role of inducible bronchus associated lymphoid tissue (iBALT) in respiratory immunity. Nat Med 10:927–934

Muniz LR, Pacer ME, Lira SA, Furtado GC (2011) A critical role for dendritic cells in the formation of lymphatic vessels within tertiary lymphoid structures. J Immunol 187:828–834

Munywoki PK, Ohuma EO, Ngama M, Bauni E, Scott JA, Nokes DJ (2013) Severe lower respiratory tract infection in early infancy and pneumonia hospitalizations among children, Kenya. Emerg Infect Dis 19:223–229

Mzinza DT, Fleige H, Laarmann K, Willenzon S, Ristenpart J, Spanier J, Sutter G, Kalinke U, Valentin-Weigand P, Förster R (2018) Application of light sheet microscopy for qualitative and quantitative analysis of bronchus-associated lymphoid tissue in mice. Cell Mol Immunol 15:875–887

Naradikian MS, Myles A, Beiting DP, Roberts KJ, Dawson L, Herati RS, Bengsch B, Linderman SL, Stelekati E, Spolski R, Wherry EJ, Hunter C, Hensley SE, Leonard WJ, Cancro MP (2016) Cutting edge: IL-4, IL-21, and IFN-γ interact to govern T-bet and CD11c expression in TLR-activated B cells. J Immunol 197:1023–1028

Neely HR, Flajnik MF (2016) Emergence and evolution of secondary lymphoid organs. Annu Rev Cell Dev Biol 32:693–711

Noble A, Zhao J (2016) Follicular helper T cells are responsible for IgE responses to Der p 1 following house dust mite sensitization in mice. Clin Exp Allergy 46:1075–1082

Orme IM, Robinson RT, Cooper AM (2015) The balance between protective and pathogenic immune responses in the TB-infected lung. Nat Immunol 16:57–63

Otsuki Y, Ito Y, Magari S (1989) Lymphocyte subpopulations in high endothelial venules and lymphatic capillaries of bronchus-associated lymphoid tissue (BALT) in the rat. Am J Anat 184:139–146

Peske JD, Thompson ED, Gemta L, Baylis RA, Fu YX, Engelhard VH (2015) Effector lymphocyte-induced lymph node-like vasculature enables naive T-cell entry into tumours and enhanced anti-tumour immunity. Nat Commun 6:7114

Pizzolla A, Oh DY, Luong S, Prickett SR, Henstridge DC, Febbraio MA, O'Hehir RE, Rolland JM, Hardy CL (2016) High fat diet inhibits dendritic cell and t cell response to allergens but does not impair inhalational respiratory tolerance. PLoS ONE 11:e0160407

Pociask DA, Scheller EV, Mandalapu S, McHugh KJ, Enelow RI, Fattman CL, Kolls JK, Alcorn JF (2013) IL-22 is essential for lung epithelial repair following influenza infection. Am J Pathol 182:1286–1296

Rangel-Moreno J, Hartson L, Navarro C, Gaxiola M, Selman M, Randall TD (2006) Inducible bronchus-associated lymphoid tissue (iBALT) in patients with pulmonary complications of rheumatoid arthritis. J Clin Invest 116:3183–3194

Rangel-Moreno J, Moyron-Quiroz JE, Hartson L, Kusser K, Randall TD (2007) Pulmonary expression of CXC chemokine ligand 13, CC chemokine ligand 19, and CC chemokine ligand 21 is essential for local immunity to influenza. Proc Natl Acad Sci USA 104:10577–10582

Rangel-Moreno J, Carragher DM, de la Luz Garcia-Hernandez M, Hwang JY, Kusser K, Hartson L, Kolls JK, Khader SA, Randall TD (2011) The development of inducible bronchus-associated lymphoid tissue depends on IL-17. Nat Immunol 12:639–646

Reed HO, Wang L, Sonett J, Chen M, Yang J, Li L, Aradi P, Jakus Z, D'Armiento J, Hancock WW, Kahn ML (2019) Lymphatic impairment leads to pulmonary tertiary lymphoid organ formation and alveolar damage. J Clin Invest 129:2514–2526

Rendon JL, Li X, Akhtar S, Choudhry MA (2013) Interleukin-22 modulates gut epithelial and immune barrier functions following acute alcohol exposure and burn injury. Shock 39:11–18

Richert LE, Harmsen AL, Rynda-Apple A, Wiley JA, Servid AE, Douglas T, Harmsen AG (2013) Inducible bronchus-associated lymphoid tissue (iBALT) synergizes with local lymph nodes during antiviral CD4+ T cell responses. Lymphat Res Biol 11:196–202

Roberts LL, Robinson CM (2014) *Mycobacterium tuberculosis* infection of human dendritic cells decreases integrin expression, adhesion and migration to chemokines. Immunology 141:39–51

Robinson PJ, Hegele RG, Schellenberg RR (1997) Allergic sensitization increases airway reactivity in guinea pigs with respiratory syncytial virus bronchiolitis. J Allergy Clin Immunol 100:492–498

Rodda LB, Bannard O, Ludewig B, Nagasawa T, Cyster JG (2015) Phenotypic and morphological properties of germinal center dark Zone Cxcl12-expressing reticular cells. J Immunol 195:4781–4791

Sanchez-Guzman D, Le Guen P, Villeret B, Sola N, Le Borgne R, Guyard A, Kemmel A, Crestani B, Sallenave JM, Garcia-Verdugo I (2019) Silver nanoparticle-adjuvanted vaccine protects against lethal influenza infection through inducing BALT and IgA-mediated mucosal immunity. Biomaterials 217:119308

Sato J, Chida K, Suda T, Sato A, Nakamura H (2000) Migratory patterns of thoracic duct lymphocytes into bronchus-associated lymphoid tissue of immunized rats. Lung 178:295–308

Sato M, Hirayama S, Matsuda Y, Wagnetz D, Hwang DM, Guan Z, Liu M, Keshavjee S (2011) Stromal activation and formation of lymphoid-like stroma in chronic lung allograft dysfunction. Transplantation 91:1398–1405

Shenoy KV, Solomides C, Cordova F, Rogers TJ, Ciccolella D, Criner GJ (2012) Low CD4/CD8 ratio in bronchus-associated lymphoid tissue is associated with lung allograft rejection. J Transplant 2012:928081

Shilling RA, Williams JW, Perera J, Berry E, Wu Q, Cummings OW, Sperling AI, Huang H (2013) Autoreactive T and B cells induce the development of bronchus-associated lymphoid tissue in the lung. Am J Respir Cell Mol Biol 48:406–414

Shilovskiy IP, Sundukova MS, Babakhin A, Gaisina AR, Maerle AV, Sergeev IV, Nikolskiy AA, Barvinckaya ED, Kovchina VI, Kudlay DA, Nikonova AA, Khaitov MR (2019) Experimental protocol for development of adjuvant-free murine chronic model of allergic asthma. J Immunol Methods 468:10–19

Shinoda K, Hirahara K, Iinuma T, Ichikawa T, Suzuki AS, Sugaya K, Tumes DJ, Yamamoto H, Hara T, Tani-Ichi S, Ikuta K, Okamoto Y, Nakayama T (2016) Thy1+ IL-7+ lymphatic endothelial cells in iBALT provide a survival niche for memory T-helper cells in allergic airway inflammation. Proc Natl Acad Sci USA 113:E2842–E2851

Shinoda K, Hirahara K, Nakayama T (2017) Maintenance of pathogenic Th2 cells in allergic disorders. Allergol Int 66:369–376

Siemeni T, Knöfel AK, Ius F, Sommer W, Salman J, Böthig D, Falk CS, Tudorache I, Haverich A, Warnecke G (2019) Transplant arteriosclerosis in humanized mice reflects chronic lung allograft dysfunction and is controlled by regulatory T cells. J Thorac Cardiovasc Surg 157:2528–2537

Slight SR, Rangel-Moreno J, Gopal R, Lin Y, Fallert Junecko BA, Mehra S, Selman M, Becerril-Villanueva E, Baquera-Heredia J, Pavon L, Kaushal D, Reinhart TA, Randall TD, Khader SA (2013) $CXCR5^{+}$ T helper cells mediate protective immunity against tuberculosis. J Clin Invest 123:712–726

Solleti SK, Srisuma S, Bhattacharya S, Rangel-Moreno J, Bijli KM, Randall TD, Rahman A, Mariani TJ (2016) Serpine2 deficiency results in lung lymphocyte accumulation and bronchus-associated lymphoid tissue formation. FASEB J 30:2615–2626

Sørensen OE, Borregaard N (2016) Neutrophil extracellular traps—the dark side of neutrophils. J Clin Invest 126:1612–1620

Stahl FR, Heller K, Halle S, Keyser KA, Busche A, Marquardt A, Wagner K, Boelter J, Bischoff Y, Kremmer E, Arens R, Messerle M, Forster R (2013) Nodular inflammatory foci are sites of T cell priming and control of murine cytomegalovirus infection in the neonatal lung. PLoS Pathog 9:e1003828

Suda T, Chida K, Hayakawa H, Imokawa S, Iwata M, Nakamura H, Sato A (1999) Development of bronchus-associated lymphoid tissue in chronic hypersensitivity pneumonitis. Chest 115:357–363

Swain SD, Meissner N, Han S, Harmsen A (2011) Pneumocystis infection in an immunocompetent host can promote collateral sensitization to respiratory antigens. Infect Immun 79:1905–1914

Tecchio C, Micheletti A, Cassatella MA (2014) Neutrophil-derived cytokines: facts beyond expression. Front Immunol 5:508

Trujillo G, Hartigan AJ, Hogaboam CM (2010) T regulatory cells and attenuated bleomycin-induced fibrosis in lungs of CCR7-/-mice. Fibrogenesis Tissue Repair 3:18

Truxova I, Kasikova L, Hensler M, Skapa P, Laco J, Pecen L, Belicova L, Praznovec I, Halaska MJ, Brtnicky T, Salkova E, Rob L, Kodet R, Goc J, Sautes-Fridman C, Fridman WH, Ryska A, Galluzzi L, Spisek R, Fucikova J (2018) Mature dendritic cells correlate with favorable immune infiltrate and improved prognosis in ovarian carcinoma patients. J Immunother Cancer 6:139

Ulrichs T, Kosmiadi GA, Trusov V, Jörg S, Pradl L, Titukhina M, Mishenko V, Gushina N, Kaufmann SH (2004) Human tuberculous granulomas induce peripheral lymphoid follicle-like structures to orchestrate local host defence in the lung. J Pathol 204:217–228

Vanessa KH, Julia MG, Wenwei L, Michelle AL, Zarina ZR, Lina LH, Sylvie A (2015) Absence of Annexin A1 impairs host adaptive immunity against Mycobacterium tuberculosis in vivo. Immunobiology 220:614–623

Venturiello SM, Verzoletti ML, Costantino SN, Forastiero MA, Roux ME (2007) Early pulmonary response in rats infected with *Trichinella spiralis*. Parasitology 134:281–288

Weinstein AM, Storkus WJ (2016) Biosynthesis and functional significance of peripheral node addressin in cancer-associated TLO. Front Immunol 7:301

Weinstein AM, Chen L, Brzana EA, Patil PR, Taylor JL, Fabian KL, Wallace CT, Jones SD, Watkins SC, Lu B, Stroncek DF, Denning TL, Fu YX, Cohen PA, Storkus WJ (2017) Tbet and IL-36γ cooperate in therapeutic DC-mediated promotion of ectopic lymphoid organogenesis in the tumor microenvironment. Oncoimmunology 6:e1322238

Weinstein AM, Giraldo NA, Petitprez F, Julie C, Lacroix L, Peschaud F, Emile JF, Marisa L, Fridman WH, Storkus WJ, Sautès-Fridman C (2019) Association of IL-36γ with tertiary lymphoid structures and inflammatory immune infiltrates in human colorectal cancer. Cancer Immunol Immunother 68:109–120
Wendland M, Willenzon S, Kocks J, Davalos-Misslitz AC, Hammerschmidt SI, Schumann K, Kremmer E, Sixt M, Hoffmeyer A, Pabst O, Forster R (2011) Lymph node T cell homeostasis relies on steady state homing of dendritic cells. Immunity 35:945–957
Wiley JA, Harmsen AG (2008) Pneumocystis infection enhances antibody-mediated resistance to a subsequent influenza infection. J Immunol 180:5613–5624
Wiley JA, Richert LE, Swain SD, Harmsen A, Barnard DL, Randall TD, Jutila M, Douglas T, Broomell C, Young M, Harmsen A (2009) Inducible bronchus-associated lymphoid tissue elicited by a protein cage nanoparticle enhances protection in mice against diverse respiratory viruses. PLoS ONE 4:e7142
Winslow GM, Papillion AM, Kenderes KJ, Levack RC (2017) CD11c+ T-bet+ memory B cells: Immune maintenance during chronic infection and inflammation? Cell Immunol 321:8–17
Yu P, Wang Y, Chin RK, Martinez-Pomares L, Gordon S, Kosco-Vibois MH, Cyster J, Fu YX (2002) B cells control the migration of a subset of dendritic cells into B cell follicles via CXC chemokine ligand 13 in a lymphotoxin-dependent fashion. J Immunol 168:5117–5123
Zhang M, Wang Z, Graner MW, Yang L, Liao M, Yang Q, Gou J, Zhu Y, Wu C, Liu H, Zhou B, Chen X (2011) B cell infiltration is associated with the increased IL-17 and IL-22 expression in the lungs of patients with tuberculosis. Cell Immunol 270:217–223
Zhang W, Zhang H, Liu S, Xia F, Kang Z, Zhang Y, Liu Y, Xiao H, Chen L, Huang C, Shen N, Xu H, Li F (2019) Excessive CD11c. Proc Natl Acad Sci USA 116:18550–18560
Zhao Y, Olonisakin TF, Xiong Z, Hulver M, Sayeed S, Yu MT, Gregory AD, Kochman EJ, Chen BB, Mallampalli RK, Sun M, Silverstein RL, Stolz DB, Shapiro SD, Ray A, Ray P, Lee JS (2015) Thrombospondin-1 restrains neutrophil granule serine protease function and regulates the innate immune response during *Klebsiella pneumoniae* infection. Mucosal Immunol 8:896–905

Chronic Inflammation in Mucosal Tissues: Barrier Integrity, Inducible Lymphoid Tissues, and Immune Surveillance

Rajrupa Chakraborty and David D. Lo

Contents

Abstract An interesting phenomenon of chronic inflammation is that the associated cytokines can simultaneously promote inflammatory cell recruitment and tissue pathology as well as tissue regeneration and development of inducible organized lymphoid tissues (tertiary lymphoid organs or TLO), demonstrating the remarkable dynamics of the immune interactions with host tissues. In mucosal tissues, chronic immune-mediated inflammation can present a mixed inflammatory pathology including neutrophil infiltrates along with the lymphocytic aggregates. The factors driving this pattern may involve effects on barrier function as well as inducible mechanisms associated with immune surveillance. The relative contribution of these factors may be important in determining the outcome, from resolution to inflammatory stalemate to progressive tissue pathology and destruction. Here, we focus on the specific impact of cytokine-driven inducible lymphoid cells and tissues

R. Chakraborty · D. D. Lo (✉)
Division of Biomedical Sciences, School of Medicine, University of California Riverside, Riverside, CA 92521, USA
e-mail: david.lo@medsch.ucr.edu

Current Topics in Microbiology and Immunology (2020) 426: 45–64
https://doi.org/10.1007/82_2020_208

Published Online: 09 May 2020

on immune surveillance at mucosal surfaces, including the induction of epithelial M cells. We propose a model of chronic intestinal inflammation to assess the relative contributions of mucosal barrier integrity, M cell transcytosis of luminal microbes, and inducible lymphoid tissues.

1 Introduction

Tissue inflammation, which is classically characterized by the recruitment of blood-borne cells from the blood into the tissues, can be categorized into two main histological patterns: acute inflammation versus chronic inflammation. Acute inflammation is characterized by short-lived recruitment of neutrophils (usually in response to the bacterial infection) with programmed resolution mediated by phagocytic macrophages and mediators such as resolvins and maresins, which alter the patterns of cell recruitment (Sansbury and Spite 2016; Ji et al. 2011; Duvall and Levy 2016; Serhan and Levy 2018). Chronic inflammation, by contrast, is a persistent accumulation of recruited cells that can display different patterns, including (but not limited to) lymphocytic infiltration, allergic inflammation with eosinophil recruitment, or granuloma formation. The chronic inflammatory response can be due to failed resolution of an acute inflammatory response as with a persistent microbial infection, or it can be driven by an adaptive immune response to repeated antigen stimulation, as in allergic asthma, or autoimmune disease.

In mucosal tissues, such as the intestine, chronic inflammation can present a mixed picture, with both neutrophil and lymphocytic infiltration in the intestinal lamina propria. The chronic nature of the inflammation would be expected to drive a protective adaptive immune response, and indeed, the development of tertiary lymphoid tissues in the intestine suggests an expansion of local active immune surveillance and adaptive immunity. However, in some disease settings, the mixed inflammatory pattern may persist, or as in the case of Crohn's disease can also show a discontinuous "skip lesion" pattern (Kleer and Appelman 2001), where normal tissue can be immediately adjacent to fully inflamed lesions. Thus, the mechanism (s) are sought to help explain the localization and progression of chronic (especially mixed) inflammatory pathology. In this discussion, we will address the interaction between acute inflammation, immune surveillance leading to adaptive immune responses, and the mechanisms promoting chronic inflammatory pathology even in the presence of an apparent robust mucosal immune response.

2 Chronic Inflammation and Tertiary Lymphoid Organs

A chronic inflammatory response at infected and damaged tissue sites promotes a persistent interaction between local immune cells and tissue-resident stromal cells. The consequences of this interaction are the accumulation of T and B cell populations in an organized fashion to form tertiary lymphoid organs (TLO) (Barone et al. 2015; Jones et al. 2016). TLOs are highly organized ectopic lymphoid follicles that develop extrinsic to the secondary lymphoid organs (SLO) such as spleen and lymph nodes (Shipman et al. 2017). Like SLO, these lymphoid structures harbor a sophisticated organization of T cell zones and B cell follicles mimicking germinal center (GC) like structures (Shi et al. 2001). Development of TLO has been associated with the pathogenesis of several autoimmune diseases including systemic lupus erythematosus, rheumatoid arthritis, multiple sclerosis, myasthenia gravis, and diabetes (Shi et al. 2001; Hsieh et al. 2011; Chang et al. 2011; Thurlings et al. 2008; Humby et al. 2009; Wengner et al. 2007; Takemura et al. 2001; Pikor et al. 2015). In other instances, TLOs are formed in the target tissues or organs of chronic infection and inflammation including the liver in hepatitis C virus infection, the lung in influenza A virus infection, the intestines in inflammatory bowel disease as well as sites of chronic allograft rejection, and even cancer (Buettner and Lochner 2016; Olivier et al. 2016; Aloisi and Pujol-Borrell 2006; Buckley et al. 2015).

3 Cell—Cytokine Interactions in TLO Formation

The formation of organized lymphoid tissues requires the coordinated interactions of a variety of cells responding to inducing cytokines; the induced stromal cells in turn produce cytokines and chemokines that help coordinate the organization and functions of the associated cells. While many of these interactions have been well documented for the development of constitutive lymphoid organs, such as lymph nodes, inflammation-induced lymphoid tissue development also needs to incorporate the influences of inflammatory cytokines that may alter the function of the induced lymphoid tissues. Here, we will describe common cytokines and stromal cells as well as the unique role of the inflammatory cytokine tumor necrosis factor (TNFα, or in this review, TNF).

Specialized lymphoid tissues like TLO appear spontaneously at sites of inflammation where LTα1β2 signaling is critical for the priming of stromal fibroblast cells, which may trigger lymphoid tissue development (Drayton et al. 2003; Kratz et al. 1996). The role of induced stromal fibroblast cells in TLO formation is critical, although our understanding of the molecular process defining induction of the different specialized cells is limited; yet, this specialization is critical in the coordinated development of the organized lymphoid tissue. Thus, specific expression of TNF family members like LIGHT and RANKL by different subsets of stromal cells has been shown to be involved in lymphoneogenesis in different models of inflammation (Schrama et al. 2001; Yu et al. 2004; Hess et al. 2012; Mueller and Hess 2012). RANKL is also known to be critical in the induction of M cells (discussed below), an important component of mucosal immune surveillance (Knoop et al. 2009; Wood et al. 2016). Besides TNF, the IL23 family member IL17 was deemed important in the activation of lung fibroblasts in the development of induced bronchus-associated lymphoid tissue (iBALT) (Fleige et al. 2014).

One theory highlights the role of recruited and resident myeloid cells at inflamed tissue sites in triggering activation of resident fibroblasts (Peduto et al. 2009; Barone et al. 2016; Peters et al. 2011; Khader et al. 2011; Rangel-Moreno et al. 2006). In this scenario, TNF released by the local myeloid cells may upregulate the expression of other cytokines and chemokines such as IL6 and BAFF on stromal fibroblasts resulting in their activation (Hardy et al. 2012; Husson et al. 2000). Circulating monocyte-derived macrophages are recruited to sites of chronic inflammation, where they produce chemokines such as CXCL13 and CXCL12 to help initiate TLO formation (Luo et al. 2019). This process is further aided by IL7-expressing stromal fibroblast cells, which influence the secretion of these chemokines in the local immune environment (Meier et al. 2007).

B cells are another cell subset recruited to sites of chronic inflammation and are a predominant lymphoid cell population in TLO (Luo et al. 2019). CXCL13 is a B cell attracting chemokine (also known as B cell attracting chemokine 1 or BCA1), which is secreted by both macrophages and induced follicular dendritic cells (FDC) and leads to B cell aggregation (Drayton et al. 2006). B cells are also potent producers of lymphotoxins (LT) (Laskov et al. 1990). In this context, resident naïve B cells and follicular DCs may resemble lymphoid tissue inducer (LTi) cells by expressing sufficient levels of $LT\alpha1\beta2$ that could then drive differentiation and maturation of mesenchymal cells (Ngo et al. 2001; Moussion and Girard 2011).

Overexpression of tumor necrosis factor (TNF) in specific tissue locations is proposed to be involved in TLO development; indeed, the phenomenon of "lymphoid neogenesis" was characterized in the context of tissue over-production of TNF (Drayton et al. 2006; Brembilla et al. 2016; Picarella et al. 1993). Although there is some level of correlation between ectopic overexpression of TNFα and formation of TLO in the periphery (Barone et al. 2016), the mechanisms driving this phenomenon are not clearly understood. In tissues with chronic inflammation, the cytokine TNFα is also commonly found as the main contributor to inflammatory pathogenesis, since it can be produced by many different cell types, including myeloid cells, fibroblasts, and lymphocytes, as a consequence of NF-κB-dependent signaling. Indeed, TNF is a nearly ubiquitous component in any inflammatory response. Consequently, anti-TNF biologics have recently become a common therapeutic in several chronic inflammatory diseases, including plaque psoriasis, inflammatory bowel disease, and rheumatoid arthritis (Monaco et al. 2015). Since TLO formation may be viewed as a beneficial expansion of mucosal immunity, there is a potential paradox if anti-TNF therapeutics may have both an inhibitory effect on TNF-mediated inflammation as well as an anti-therapeutic effect if it also inhibits TLO formation.

Thus, TLO development can be broadly viewed as a progression beginning with signals from pro-inflammatory cytokines produced by recruited and tissue-resident myeloid cells, which trigger priming of the local stromal fibroblasts. Activation of fibroblasts, which marks the first step in TLO development, leads to further production of pro-inflammatory cytokines and adhesion molecules. Simultaneously, a variety of resident cells are induced to produce inflammatory cytokines such as TNFα, LTα, and LTα1β2, which further promotes the maturation of TLO. Fulfillment of this second step requires combined and continued activation of TNFR1 and LTβR signaling pathways (Gräbner et al. 2009; Lötzer et al. 2010) followed by downstream expression of cytokines and chemokines that systematically regulate differentiation of resident fibroblasts and give rise to specialized lymphoid tissues.

4 Customized Stromal Cells and Immune Surveillance: Intestinal M Cells

The induction of specialized lymphoid tissue stromal cells and organized structures is strongly dependent on the local tissue environment. For example, the induction of organized lymphoid tissues in pancreatic islets shows a strong resemblance to the lymph node. By contrast, expression of the lymphocyte chemokine CCL21 in the central nervous system only produces disorganized lymphocyte recruitment under specific conditions such as toxoplasma infection, and even under those conditions, organized lymphoid structures fail to appear. Expression of TNF in the CNS astrocytes induced perivascular lymphocytic infiltrates with HEV-like vascular changes, but without clear organization into lymphoid tissue (Stalder et al. 1998).

Regarding the influence of tissue environment, it should also be noted that chronic inflammation and TNF are not only associated with induction of specialized stromal cells, and it can also drive tissue stem cells or other progenitor cells promoting tissue regeneration. For example, in multiple sclerosis, TNF produced by autoimmune inflammation also drives the activation of astrocytes and remyelination by oligodendrocytes (Arnett et al. 2001; Madsen et al. 2016; Patel et al. 2012). The intestine may have parallels to this phenomenon, both in the induction of the organized lymphoid stromal cells, but perhaps also in the regulation of the epithelial barrier.

Returning to our focus on chronic inflammation in the intestine, here too the tissue determines local customization in the induction of specialized stromal cells. Thus, the unique feature of mucosal lymphoid tissue is that in contrast to lymph nodes which rely on lymphatic drainage of myeloid antigen-presenting cells (e.g., Langerhans cells, dendritic cells) from the tissue into the lymph node, antigen delivery is instead provided only at the luminal epithelium by M cells. M cells are crypt stem cell-derived epithelial cells that have acquired specialized ability to capture microparticles from the intestinal lumen and transport them across the epithelial barrier to waiting dendritic cells in the organized lymphoid tissues (e.g., Peyer's patches) below (Kraehenbuhl and Neutra 2000; Lo 2017; Dillon and Lo 2019). Their induction is largely dependent on expression of RANKL by induced reticular fibroblastic cells, normally found among stromal cells in the organized lymphoid tissues, though in vitro and in vivo studies support a role for TNF as well in M cell induction (Knoop et al. 2009; Wood et al. 2016).

The contrast with immune surveillance mechanisms in solid organs is intriguing; in most tissues, resident myeloid cells such as tissue macrophages and skin Langerhans cells use scavenger receptors and phagocytosis to acquire antigen, and then directly present processed antigen to lymphocytes to initiate immune responses. M cells, which are non-motile epithelial cells, instead use a combination of mechanisms including changes at their apical surface to enable capture of suspended biological microparticles based on surface charge. A few surface molecules have also been identified as "capture receptors" for pathogen entry (Lo et al. 2012; Hase et al. 2009; Neutra et al. 1996a, b; Barton et al. 2001; Clark et al. 1998; Mengaud et al. 1996), though it could be argued that in many cases this mechanism was a pathogen adaptation to the host rather than an evolved M cell-specific strategy for general immune surveillance.

M cells are most commonly found in the epithelium overlying organized mucosal lymphoid tissues such as Peyer's patches and ILFs, and this characteristic localization highlights the role M cells play in mucosal immune surveillance. However, because they are epithelial cells, they are only specialized in microparticle capture and do not perform antigen processing or presentation, and do not express any lymphocyte co-stimulation ligands. Instead, they transport the particles and antigens across the epithelial barrier for delivery to intimately associated sub-epithelial dendritic cells; thus, they act principally as "antigen delivery" cells rather than "antigen-presenting" cells. Separating the delivery from presentation among two different cell types at the epithelial barrier allows a neat differentiation of specialized lineages, but it also separates functional penetration of the barrier from functional immune surveillance (Fig. 1).

Mucosal lymphoid tissues and M cells appear to have an interesting relationship to chronic inflammation and TNF. While Peyer's patches and M cells are found in mice lacking either of the TNF receptor genes (Parnell et al. 2017), these might be categorized as "constitutive" mucosal lymphoid tissues, to be contrasted with "inflammation-inducible" tissues (Lo 2017; Dillon and Lo 2019). In studies on chronic intestinal inflammation in mice induced by DSS, M cells were induced throughout the intestinal epithelium, but these were distinct from constitutive M cells, as treatment with anti-TNF antibodies abrogated their induction (Bennett et al. 2016). Interestingly, the widespread inflammatory response in this model led to the recruitment of lymphocytes into lamina propria aggregates, but the M cell induction was not strictly dependent on the induction of other stromal cells, suggesting that inflammation-induced M cells can appear independently of lymphoid organ stromal cells. This disconnect may have important consequences discussed later.

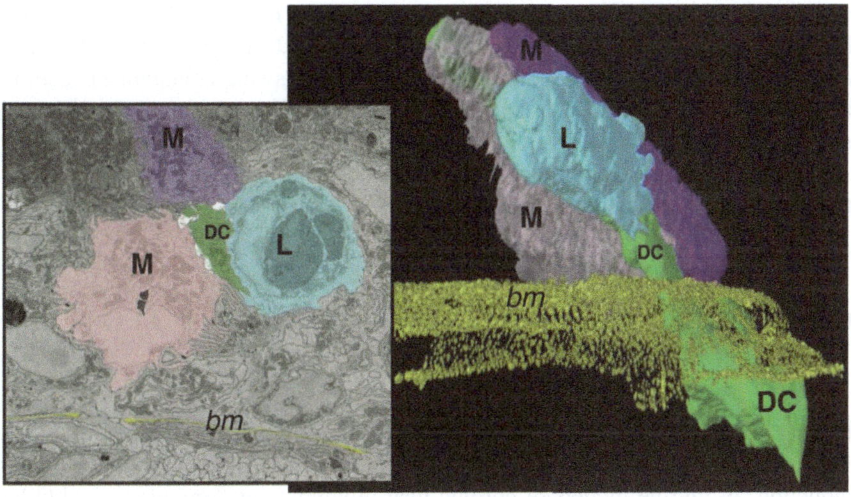

Fig. 1 Illustration of the specialization of M cell "antigen delivery" at the epithelial barrier versus dendritic cell "antigen presentation" below the epithelial barrier. (Left) Electron microscopy image showing the close relationship between dendritic cell process (green) and M cells (pink and purple) above the epithelial basement membrane (bm, yellow). Lymphocyte (blue), possibly a B cell, is shown adjacent to the M cells. (Right) Serial images of the same preparation were traced and assembled into a 3D projection, showing the dendritic cell process reaching to interact with M cells, while the cell body remains below the basement membrane (yellow) (Justin Chen)

5 Airways: Chronic Inflammation and Bronchus-Associated Lymphoid Tissues

The discussion so far has focused on the intestinal epithelial barrier, but related issues are present in the airway, where immune surveillance at the epithelial barrier shows a similar organization of organized lymphoid tissues. In the upper airway, nasopharyngeal-associated lymphoid tissue (NALT) (Zuercher et al. 2002; Claeys et al. 1996; Hameleers et al. 1989; Harmsen et al. 2002; Wu et al. 1997) in the mouse and tonsil in humans both depend on epithelial M cells for antigen and microparticle capture for delivery to underlying immune tissue. Deeper in the lung, bronchus-associated lymphoid tissue (BALT) is also found in the sub-epithelial tissue of the larger airways, with the organization similar to intestinal Peyer's patch, including epithelial M cells (Tango et al. 2000; Kim et al. 2011).

The impact of chronic inflammation on the epithelial barrier and organized lymphoid tissue is also similar, though chronic inflammation here follows different patterns. For example, allergic inflammation produces a picture of chronic inflammation and lymphoid aggregates in the sub-epithelial tissues, accompanied by lymphocyte, eosinophil, and neutrophil recruitment in the alveolar compartment. In mice, organized lymphoid tissues such as BALT are less commonly found under common vivarium conditions, though in the presence of chronic inflammation and advanced age these tissues are more easily identified. Epithelial M cells have also been identified, at least in the BALT of aging animals (Tango et al. 2000). Repeated administration of aerosols of agricultural dusts from swine containment facilities also has been shown to promote both neutrophil recruitment accompanied by sub-epithelial lymphocyte aggregates (Poole et al. 2011, 2012, 2017), with some similarities to organized lymphoid tissues. The contribution of TNF is not well characterized in these situations.

6 Inducible Lymphoid Structure in the Intestine: TLO Versus ILF

Having now covered the basic components in intestinal inducible lymphoid tissues, we now discuss the potential functional implications of this lymphoid tissue induction. In murine models of chronic ileal inflammation, formation of TLO has been reported in the mesentery (Rehal and von der Weid 2017). Additionally, it was proposed that formation of these mesenteric TLO was positively correlated with disease progression, which was further supported by the observation of similar lymphoid aggregates in the ileal mesentery of patients with Crohn's disease (Randolph et al. 2016). An intriguing theory is that chronic inflammation and submucosal edema contribute to leaky lymphatic vessels allowing drainage of immune cells and antigens in the adjoining tissue spaces, thereby promoting lymphoid neogenesis.

In many ways, TLOs are similar as well as distinct from the intestinal-isolated lymphoid follicles (ILFs). (1) Genesis and maturation of ILFs in the gut require commensal recognition by the toll-like receptors as well as RORγt-expressing LTi cells, found in the crypts of the intestinal lamina propria (Bouskra et al. 2008; Tsuji et al. 2008; Eberl et al. 2004; Eberl and Littman 2004). Consequently, maturation of ILFs seems to be impaired in germ-free mice (Hamada et al. 2002; Pabst et al. 2006). Although persistent influx of antigens seems to be important for TLO formation, it is not yet clear whether microbial antigens are necessary and sufficient for the development of TLO. (2) The immune environments of TLO and ILFs are also distinct, whereas TLOs harbor a consortium of lymphoid and myeloid cells, ILFs, on the other hand, primarily constitute organized B cell follicles (Lorenz et al. 2003). (3) Evidence suggests that a reciprocal relationship between gut bacteria and ILFs is essential for the regulation and maintenance of intestinal homeostasis (Bouskra et al. 2008). Whether development of TLO is associated with resolution or promotion of inflammation, however, remains to be determined.

7 TLO: Are They Protective?

Recent studies have underscored the beneficial roles of TLO in microbial infection and inflammation. Formation of TLO at ectopic locations can trigger antigen-specific immune responses mediated primarily by plasma B cells and antibodies in response to microbial and inflammatory stimuli (Schröder et al. 1996; Dörner et al. 2002). In some microbial infections, TLOs serve as local "powerhouses" of immune activities, which is supported by an inflammatory milieu (Ghosh et al. 2005). These local immune responses play an important role in restricting microbial dissemination and may also contribute to their clearance (Ghosh et al. 2005; Steere et al. 1988). Lung TLO is a positive development during influenza virus infection where strong antigen-specific T cell responses drive viral clearance and maintain tissue tolerance (Moyron-Quiroz et al. 2004, 2006). Additionally, lymphocyte exit is believed to be an important phenomenon during TLO-associated lymphangiogenesis insinuating a pro-resolving mechanism of TLO in chronic inflammation (Nayar et al. 2016). Considering the important role of lymphocyte-derived LTα1β2 in TLO-associated lymphangiogenesis, it has been used as a therapeutic target for diseases like rheumatoid arthritis (Gommerman and Browning 2003; Gatumu et al. 2009; Wu et al. 2001). However, blocking LTα1β2 signaling pathway in TLO-associated diseases may impair complete remodeling of lymphatic vessels and subsequently lymphocyte egress, thereby impacting resolution of inflammation and regulation of tissue tolerance. Based on this evidence, it is conceivable that inflammation-associated mesenteric TLO may play a protective role to ameliorate or even resolve inflammation.

Current evidence supporting the protective role of TLO is limited and to some extent controversial. Characterization of the immune repertoires in mature TLO revealed the presence of regulatory T cells (Tregs) as well as a heterogeneous population of B cells that could participate in immunosuppression (Yin et al. 2016). Since follicular DCs in the TLO serve as long-term storehouses of microbial antigens, they may influence antigen-specific B cell maturation including GC reactions (somatic hypermutation and affinity maturation). Indeed, class-switched B cells specifically CD19+/IgG1+ and CD19+/IgA+ B2 cells were observed in some TLO, which bears evidence to matured GC reactions and supports the above hypothesis (Srikakulapu et al. 2016). In this context, production of regulatory B cell cytokines like IL10, TGFβ1 by B1 cells may aid in the regulation of GC reactions, B cell homeostasis, and eventually bolster immunosuppressive functions (Lykken et al. 2015; Mauri and Menon 2015; Miyagaki 2015; Ray et al. 2012; Tedder 2015; Shen and Fillatreau 2015).

The T cell repertoire in TLO includes both natural and induced Tregs, which may have important immunosuppressive functions in the context of chronic inflammatory diseases (Hu et al. 2015). However, these cells have been reported to play controversial roles in some models of chronic inflammation, for example atherosclerosis (Hu et al. 2015; Clement et al. 2015), thereby stressing the need to study the roles of these cells in the regulation of T cell immunity in TLO-associated diseases. Naïve T cells are rare in TLO, but it could be hypothesized that they differentiate into regulatory T memory cells in chronic inflammatory settings, thereby contributing to antigen-specific immunosuppression (Bilate and Lafaille 2012; Curotto de Lafaille and Lafaille 2009). Adoptive transfer studies in aged ApoE-/- mice revealed that Tregs in the TLO may be generated locally via clonal selection of endogenous Tregs in an antigen-dependent manner (Hu et al. 2015). It is possible that these processes are dictated by several other factors including the age of mice and even the local immune and inflammatory environment. Further studies suggested an important role of naïve CD4+ T cells in the generation of TLO-specific induced Tregs (Hu et al. 2015); however, the functional significance of these regulatory T cells in chronic inflammation is not fully understood. It appears that a threshold exists, governed by the availability of antigens, inflammatory signals, and regulation of homeostasis in the T cell and B cell compartments. Chronic inflammation may tip this balance from immunosuppression toward sustained immune activation, thereby contributing to disease exacerbation.

Interestingly, not all diseases with chronic inflammation have been associated with TLO formation but this development seems to be restricted to "permissive" tissue sites, one classic example being the mucosal epithelium (Barone et al. 2016). Although factors defining permissive tissue locations are currently unclear, persistent exposure to antigens from the intestinal microbiota, in addition to baseline inflammatory and other immune stimuli, could promote TLO formation.

It could be hypothesized that, in most settings, the development of TLO is a response to, and not necessarily a cause of, inflammation. Tissue damage in conjunction with defects in mucosal barrier integrity may result in increased flux of luminal antigens across the epithelium that can initiate a pro-inflammatory cascade. Enhanced local concentrations of cytokines and chemokines may induce accumulation of both myeloid and lymphoid cell populations, which under the influence of appropriate stimuli in the microenvironment may form organized immune cell aggregates promoting lymphoid neogenesis. Indeed, it is important to determine the factors triggering TLO formation; however, the mere presence of these lymphoid aggregates does not guarantee a positive correlation with disease progression and clinical outcomes.

8 Mucosal Epithelium Barrier Function and Immune Surveillance

We now turn to the question of chronic inflammation in the intestine and its impact on barrier function and immune surveillance. In contrast to solid organs, chronic inflammation at the intestinal barrier presents different challenges. Here, the presence of intestinal luminal microbiota illustrates the importance of "location, location, location" in health and disease. Microbes in the intestinal lumen may be part of a balanced microbial ecology and include bacteria that would be considered "commensals" in the right context. However, the happy relationship is dependent on the principle of "fences make for good neighbors," as most bacteria are welcome as long as they stay in the lumen. Breaches in the intestinal barrier threaten the balance by triggering innate immune responses and inflammation. In this context, there are three aspects that are critical to both barrier function and immune surveillance: (1) intestinal epithelium barrier integrity, (2) surveillance across the barrier by resident dendritic cells, and (3) M cell transcytosis across the barrier.

Inflammation and the cytokines produced are threats to the epithelial barrier not just from potential cytolytic effects, but also to the integrity of the tight junction. Extensive studies have identified a number of ways in which tight junctions can be altered by inflammatory cytokines such as interferon gamma (IFN-γ) (Chiba et al. 2006) and TNF (Marchiando et al. 2010), including changes in tight junction components (e.g., changes in claudin gene expression) and changes in the distribution of tight junction proteins, including recycling of proteins away from the tight junction (Odenwald and Turner 2013; Van Itallie and Anderson 2014).

Intestinal epithelium barrier integrity at the tight junction is influenced by other factors that are not always associated with chronic inflammation. For example, luminal microbes also promote tight junction integrity through direct action on intestinal epithelium, signaling through TLR2. In gnotobiotic mice, intestinal barrier function is decreased due to the paucity of this signal, which actually promotes the expression of tight junction proteins by the epithelium (Cario et al. 2007; Cario 2008).

Other signals within intestinal epithelium are also influential in maintaining barrier function. Mutations in PTPN2 are linked to susceptibility to IBD, and studies in mice suggest that a loss-of-function mutation in this phosphatase would result in enhanced JAK-STAT signaling, potentiating the effects of pro-inflammatory cytokines such as IFN-γ on epithelial tight junctions (Spalinger et al. 2016, 2018).

How would defects in barrier integrity affect chronic inflammation and pathogenesis? The tight junction defects, ranging from altered claudin gene expression to loss of tight junction components, can lead to increased permeability to ions, larger molecules (assayed by permeability to dextran-FITC), or even globular proteins (Van Itallie and Anderson 2014; Van Itallie et al. 2008). Within this size range, innate immune ligands such as endotoxin and peptidoglycan can cross the barrier and potentially trigger tissue inflammatory signals. Molecules in this range can also be carried across the epithelial barrier by the pores provided by goblet cells, though the relative contribution of this pathway would depend on the severity of the tight junction defect.

9 Mucosal Barrier Versus Protective Immunity—Barriers, Immunity, and Inflammation

Given the microbiota present in the intestinal lumen, one might expect that microbial components leaking across tight junctions or goblet cell pores would lead to potential for innate immune-driven chronic inflammation. Yet under steady-state conditions, this does not appear to be the case. Is there a robust protective immunity to prevent the inflammatory response, a low-level tolerance to leaking innate immune ligands, another unidentified mechanism, or is this leak pathway simply insufficient to promote chronic inflammatory pathogenesis?

If there is a protective immune effector (or suppressor/regulatory) mechanism, immune surveillance would potentially be required. In the setting of barrier function defects, this surveillance would in principle depend mainly on resident macrophages in the lamina propria, but due to cell migration pathways, these macrophages are more likely to enter draining mesenteric lymphoid tissues rather than lamina propria Peyer's patches, TLO, or ILF. How then would a protective immune response augment epithelial barrier function?

We referred to the presence of organized lymphoid tissue in mucosal tissues such as intestine (and lung) as critical for the development of protective immune responses, especially secretory IgA, though in some cases mucosal immunity can develop in the apparent absence of organized mucosal immune tissue. In the setting of chronic inflammation, the generation of organized lymphoid tissue is enhanced, which would presumably also enhance protective immunity, reducing inflammation-related pathogenesis and disease progression. However, as discussed above, inflammation also has a deleterious effect on overall epithelial barrier integrity.

In an attempt to make sense of all of these factors, we propose a scenario that may help clarify the relationship between epithelial barrier function and induction of organized lymphoid tissues in the setting of chronic inflammation. The initial trigger of inflammation may be due to any of several mechanisms, including microbial infection, immune effector activity, or breaches in the epithelial barrier (e.g., ulceration), so this scenario seeks an explanation for the maintenance or progression of inflammatory pathology.

We propose three parallel mechanisms (Fig. 2), each with different impacts on the maintenance of chronic inflammation and pathogenesis. (1) At one end, we view the inflammatory effects on epithelial barrier function to result in increased permeability to innate immune ligands, triggering induction of cytokines such as TNF and IL-1β, but not crossing the threshold to trigger broad recruitment of inflammatory cells. (2) Inflammatory cytokines, such as TNF, induce changes in crypt stem cells promoting the production of M cells, but asynchronous with the formation of mature organized lymphoid tissues; transcytosis of microbes from the lumen into lamina propria would drive innate immune signals and promote neutrophil recruitment. (3) Inflammatory cytokine/chemokine production by resident

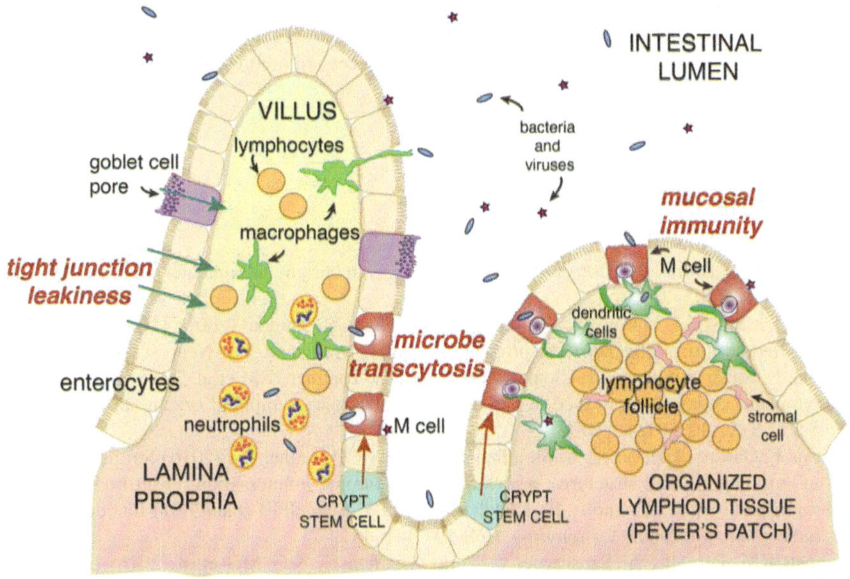

Fig. 2 Mucosal barrier, immune surveillance, and inflammation. Model showing three factors affecting intestinal chronic inflammation. Tight junction leakiness and M cell microbe transcytosis may be factors contributing to ongoing chronic inflammation; induction of organized lymphoid tissue may be a product of the inflammation, but without direct impact on the first two factors

cells leads to the recruitment of lymphocytes into the lamina propria aggregates, eventually leading to the formation of organized lymphoid tissues.

Note that this scenario provides a few interacting mechanisms that would potentially reinforce the inflammatory pathogenesis. Barrier permeability would lead to inflammatory cytokine-mediated induction of stem cells to produce M cells. M cell in turn would transcytose luminal bacteria into the lamina propria. Since bacteria could enter tissue in the absence of organized lymphoid tissues to capture them, they would drive further innate immune pro-inflammatory responses, including neutrophil recruitment and macrophage activation. This inflammatory response would in turn drive further production of cytokines to degrade barrier function and reinforce the signals driving M cell production. Lymphocyte recruitment and formation of organized lymphoid tissues should lead to enhanced immune surveillance and production of protective immunity, including secretory IgA responses. However, the production of IgA may be of little impact on the effect of microbial transcytosis into lamina propria, and the chronic inflammatory pattern is sustained. Resolution of the inflammatory response may depend on sufficient recruitment of neutrophils and macrophages to sterilize the lamina propria, as well as production and recruitment of regulatory T cells.

Acknowledgements This work was supported by the National Institute of Allergy and Infectious Diseases of the National Institutes of Health under award number R01/R56 AI63426. The content is solely the responsibility of the authors and does not necessarily represent the official views of the National Institutes of Health.

Bibliography

Aloisi F, Pujol-Borrell R (2006) Lymphoid neogenesis in chronic inflammatory diseases. Nat Rev Immunol 6:205–217

Arnett HA, Mason J, Marino M, Suzuki K, Matsushima GK, Ting JP (2001) TNF alpha promotes proliferation of oligodendrocyte progenitors and remyelination. Nat Neurosci 4:1116–1122

Barone F, Nayar S, Campos J, Cloake T, Withers DR, Toellner K-M et al (2015) IL-22 regulates lymphoid chemokine production and assembly of tertiary lymphoid organs. Proc Natl Acad Sci U S A 112:11024–11029

Barone F, Gardner DH, Nayar S, Steinthal N, Buckley CD, Luther SA (2016) Stromal fibroblasts in tertiary lymphoid structures: a novel target in chronic inflammation. Front Immunol 7:477

Barton ES, Forrest JC, Connolly JL, Chappell JD, Liu Y, Schnell FJ et al (2001) Junction adhesion molecule is a receptor for reovirus. Cell 104:441–451

Bennett KM, Parnell EA, Sanscartier C, Parks S, Chen G, Nair MG et al (2016) Induction of colonic M cells during intestinal inflammation. Am J Pathol 186:1166–1179

Bilate AM, Lafaille JJ (2012) Induced CD4+ Foxp3+ regulatory T cells in immune tolerance. Annu Rev Immunol 30:733–758

Bouskra D, Brézillon C, Bérard M, Werts C, Varona R, Boneca IG et al (2008) Lymphoid tissue genesis induced by commensals through NOD1 regulates intestinal homeostasis. Nature 456:507–510

Brembilla NC, Dufour AM, Alvarez M, Hugues S, Montanari E, Truchetet M-E et al (2016) IL-22 capacitates dermal fibroblast responses to TNF in scleroderma. Ann Rheum Dis 75:1697–1705

Buckley CD, Barone F, Nayar S, Bénézech C, Caamaño J (2015) Stromal cells in chronic inflammation and tertiary lymphoid organ formation. Annu Rev Immunol 33:715–745

Buettner M, Lochner M (2016) Development and function of secondary and tertiary lymphoid organs in the small intestine and the colon. Front Immunol 7:342

Cario E (2008) Barrier-protective function of intestinal epithelial Toll-like receptor 2. Mucosal Immunol 1(Suppl 1):S62–S66

Cario E, Gerken G, Podolsky DK (2007) Toll-like receptor 2 controls mucosal inflammation by regulating epithelial barrier function. Gastroenterology 132:1359–1374

Chang A, Henderson SG, Brandt D, Liu N, Guttikonda R, Hsieh C et al (2011) In situ B cell-mediated immune responses and tubulointerstitial inflammation in human lupus nephritis. J Immunol Baltim Md 1950 186:1849–1860

Chiba H, Kojima T, Osanai M, Sawada N (2006) The significance of interferon-gamma-triggered internalization of tight-junction proteins in inflammatory bowel disease. Sci STKE Signal Transduct Knowl Environ 2006:pe1

Claeys S, Cuvelier C, Quatacker J, Van Cauwenberge P (1996) Ultrastructural investigation of M-cells and lymphoepithelial contacts in naso-pharyngeal associated lymphoid tissue (NALT). Acta Oto-Laryngol Suppl 523:40–42

Clark MA, Hirst BH, Jepson MA (1998) M-cell surface beta1 integrin expression and invasin-mediated targeting of *Yersinia pseudotuberculosis* to mouse Peyer's patch M cells. Infect Immun 66:1237–1243

Clement M, Guedj K, Andreata F, Morvan M, Bey L, Khallou-Laschet J et al (2015) Control of the T follicular helper-germinal center B-cell axis by $CD8^+$ regulatory T cells limits atherosclerosis and tertiary lymphoid organ development. Circulation 131:560–570

Curotto de Lafaille MA, Lafaille JJ (2009) Natural and adaptive foxp3+ regulatory T cells: more of the same or a division of labor? Immunity 30:626–635

Dillon A, Lo DD (2019) M Cells: intelligent engineering of mucosal immune surveillance. Front Immunol 10:1499

Dörner T, Hansen A, Jacobi A, Lipsky PE (2002) Immunglobulin repertoire analysis provides new insights into the immunopathogenesis of Sjögren's syndrome. Autoimmun Rev 1:119–124

Drayton DL, Ying X, Lee J, Lesslauer W, Ruddle NH (2003) Ectopic LT alpha beta directs lymphoid organ neogenesis with concomitant expression of peripheral node addressin and a HEV-restricted sulfotransferase. J Exp Med 197:1153–1163

Drayton DL, Liao S, Mounzer RH, Ruddle NH (2006) Lymphoid organ development: from ontogeny to neogenesis. Nat Immunol 7:344–353

Duvall MG, Levy BD (2016) DHA- and EPA-derived resolvins, protectins, and maresins in airway inflammation. Eur J Pharmacol 785:144–155

Eberl G, Littman DR (2004) Thymic origin of intestinal alphabeta T cells revealed by fate mapping of RORgammat+ cells. Science 305:248–251

Eberl G, Marmon S, Sunshine M-J, Rennert PD, Choi Y, Littman DR (2004) An essential function for the nuclear receptor RORgamma(t) in the generation of fetal lymphoid tissue inducer cells. Nat Immunol 5:64–73

Fleige H, Ravens S, Moschovakis GL, Bölter J, Willenzon S, Sutter G et al (2014) IL-17-induced CXCL12 recruits B cells and induces follicle formation in BALT in the absence of differentiated FDCs. J Exp Med 211:643–651

Gatumu MK, Skarstein K, Papandile A, Browning JL, Fava RA, Bolstad AI (2009) Blockade of lymphotoxin-beta receptor signaling reduces aspects of Sjögren's syndrome in salivary glands of non-obese diabetic mice. Arthritis Res Ther 11:R24

Ghosh S, Steere AC, Stollar BD, Huber BT (2005) In situ diversification of the antibody repertoire in chronic Lyme arthritis synovium. J Immunol Baltim Md 1950 174:2860–2869

Gommerman JL, Browning JL (2003) Lymphotoxin/light, lymphoid microenvironments and autoimmune disease. Nat Rev Immunol 3:642–655

Gräbner R, Lötzer K, Döpping S, Hildner M, Radke D, Beer M et al (2009) Lymphotoxin beta receptor signaling promotes tertiary lymphoid organogenesis in the aorta adventitia of aged ApoE-/- mice. J Exp Med 206:233–248

Hamada H, Hiroi T, Nishiyama Y, Takahashi H, Masunaga Y, Hachimura S et al (2002) Identification of multiple isolated lymphoid follicles on the antimesenteric wall of the mouse small intestine. J Immunol Baltim Md 1950 168:57–64

Hameleers DM, van der Ende M, Biewenga J, Sminia T (1989) An immunohistochemical study on the postnatal development of rat nasal-associated lymphoid tissue (NALT). Cell Tissue Res 256:431–438

Hardy R, Juarez M, Naylor A, Tu J, Rabbitt EH, Filer A et al (2012) Synovial DKK1 expression is regulated by local glucocorticoid metabolism in inflammatory arthritis. Arthritis Res Ther 14: R226

Harmsen A, Kusser K, Hartson L, Tighe M, Sunshine MJ, Sedgwick JD et al (2002) Cutting edge: organogenesis of nasal-associated lymphoid tissue (NALT) occurs independently of lymphotoxin-alpha (LT alpha) and retinoic acid receptor-related orphan receptor-gamma, but the organization of NALT is LT alpha dependent. J Immunol Baltim Md 1950 168:986–990

Hase K, Kawano K, Nochi T, Pontes GS, Fukuda S, Ebisawa M et al (2009) Uptake through glycoprotein 2 of FimH(+) bacteria by M cells initiates mucosal immune response. Nature 462:226–230

Hess E, Duheron V, Decossas M, Lézot F, Berdal A, Chea S et al (2012) RANKL induces organized lymph node growth by stromal cell proliferation. J Immunol Baltim Md 1950 188:1245–1254

Hsieh C, Chang A, Brandt D, Guttikonda R, Utset TO, Clark MR (2011) Predicting outcomes of lupus nephritis with tubulointerstitial inflammation and scarring. Arthritis Care Res 63:865–874

Hu D, Mohanta SK, Yin C, Peng L, Ma Z, Srikakulapu P et al (2015) Artery tertiary lymphoid organs control aorta immunity and protect against atherosclerosis via vascular smooth muscle cell lymphotoxin β receptors. Immunity 42:1100–1115

Humby F, Bombardieri M, Manzo A, Kelly S, Blades MC, Kirkham B et al (2009) Ectopic lymphoid structures support ongoing production of class-switched autoantibodies in rheumatoid synovium. PLoS Med 6:e1

Husson H, Lugli SM, Ghia P, Cardoso A, Roth A, Brohmi K et al (2000) Functional effects of TNF and lymphotoxin alpha1beta2 on FDC-like cells. Cell Immunol 203:134–143

Ji R-R, Xu Z-Z, Strichartz G, Serhan CN (2011) Emerging roles of resolvins in the resolution of inflammation and pain. Trends Neurosci 34:599–609

Jones GW, Hill DG, Jones SA (2016) Understanding immune cells in tertiary lymphoid organ development: it is all starting to come together. Front Immunol 7:401

Khader SA, Guglani L, Rangel-Moreno J, Gopal R, Junecko BAF, Fountain JJ et al (2011) IL-23 is required for long-term control of Mycobacterium tuberculosis and B cell follicle formation in the infected lung. J Immunol Baltim Md 1950 187:5402–5407

Kim D-Y, Sato A, Fukuyama S, Sagara H, Nagatake T, Kong IG et al (2011) The airway antigen sampling system: respiratory m cells as an alternative gateway for inhaled antigens. J Immunol Baltim Md 1950 186:4253–4262

Kleer CG, Appelman HD (2001) Surgical pathology of Crohn's disease. Surg Clin North Am 81:13–30

Knoop KA, Kumar N, Butler BR, Sakthivel SK, Taylor RT, Nochi T et al (2009) RANKL is necessary and sufficient to initiate development of antigen-sampling M cells in the intestinal epithelium. J Immunol Baltim Md 1950 183:5738–5747

Kraehenbuhl JP, Neutra MR (2000) Epithelial M cells: differentiation and function. Annu Rev Cell Dev Biol 16:301–332

Kratz A, Campos-Neto A, Hanson MS, Ruddle NH (1996) Chronic inflammation caused by lymphotoxin is lymphoid neogenesis. J Exp Med 183:1461–1472

Laskov R, Lancz G, Ruddle NH, McGrath KM, Specter S, Klein T et al (1990) Production of tumor necrosis factor (TNF-alpha) and lymphotoxin (TNF-beta) by murine pre-B and B cell lymphomas. J Immunol Baltim Md 1950 144:3424–3430

Lo DD (2017) Vigilance or subversion? Constitutive and inducible M cells in mucosal tissues. Trends Immunol

Lo DD, Ling J, Eckelhoefer AH (2012) M cell targeting by a Claudin 4 targeting peptide can enhance mucosal IgA responses. BMC Biotechnol 12:7

Lorenz RG, Chaplin DD, McDonald KG, McDonough JS, Newberry RD (2003) Isolated lymphoid follicle formation is inducible and dependent upon lymphotoxin-sufficient B lymphocytes, lymphotoxin beta receptor, and TNF receptor I function. J Immunol Baltim Md 1950 170:5475–5482

Lötzer K, Döpping S, Connert S, Gräbner R, Spanbroek R, Lemser B et al (2010) Mouse aorta smooth muscle cells differentiate into lymphoid tissue organizer-like cells on combined tumor necrosis factor receptor-1/lymphotoxin beta-receptor NF-kappaB signaling. Arterioscler Thromb Vasc Biol 30:395–402

Luo S, Zhu R, Yu T, Fan H, Hu Y, Mohanta SK et al (2019) Chronic inflammation: a common promoter in tertiary lymphoid organ neogenesis. Front Immunol 10:2938

Lykken JM, Candando KM, Tedder TF (2015) Regulatory B10 cell development and function. Int Immunol 27:471–477

Madsen PM, Motti D, Karmally S, Szymkowski DE, Lambertsen KL, Bethea JR et al (2016) Oligodendroglial TNFR2 mediates membrane TNF-dependent repair in experimental autoimmune encephalomyelitis by promoting oligodendrocyte differentiation and remyelination. J Neurosci Off J Soc Neurosci 36:5128–5143

Marchiando AM, Shen L, Graham WV, Weber CR, Schwarz BT, Austin JR et al (2010) Caveolin-1–dependent occludin endocytosis is required for TNF-induced tight junction regulation in vivo. J Cell Biol 189:111–126

Mauri C, Menon M (2015) The expanding family of regulatory B cells. Int Immunol 27:479–486

Meier D, Bornmann C, Chappaz S, Schmutz S, Otten LA, Ceredig R et al (2007) Ectopic lymphoid-organ development occurs through interleukin 7-mediated enhanced survival of lymphoid-tissue-inducer cells. Immunity 26:643–654

Mengaud J, Ohayon H, Gounon P, Mege R-M, Cossart P (1996) E-cadherin is the receptor for internalin, a surface protein required for entry of *L. monocytogenes* into epithelial cells. Cell 84:923–932

Miyagaki T (2015) Regulatory B cells in human autoimmune diseases. Nihon Rinsho Meneki Gakkai Kaishi 38:390–397

Monaco C, Nanchahal J, Taylor P, Feldmann M (2015) Anti-TNF therapy: past, present and future. Int Immunol 27:55–62

Moussion C, Girard J-P (2011) Dendritic cells control lymphocyte entry to lymph nodes through high endothelial venules. Nature 479:542–546

Moyron-Quiroz JE, Rangel-Moreno J, Kusser K, Hartson L, Sprague F, Goodrich S et al (2004) Role of inducible bronchus associated lymphoid tissue (iBALT) in respiratory immunity. Nat Med 10:927–934

Moyron-Quiroz JE, Rangel-Moreno J, Hartson L, Kusser K, Tighe MP, Klonowski KD et al (2006) Persistence and responsiveness of immunologic memory in the absence of secondary lymphoid organs. Immunity 25:643–654

Mueller CG, Hess E (2012) Emerging Functions of RANKL in Lymphoid Tissues. Front Immunol 3:261

Nayar S, Campos J, Chung MM, Navarro-Núñez L, Chachlani M, Steinthal N et al (2016) Bimodal expansion of the lymphatic vessels is regulated by the sequential expression of IL-7 and lymphotoxin α1β2 in newly formed tertiary lymphoid structures. J Immunol Baltim Md 1950 197:1957–1967

Neutra MR, Frey A, Kraehenbuhl JP (1996a) Epithelial M cells: gateways for mucosal infection and immunization. Cell 86:345–348

Neutra MR, Pringault E, Kraehenbuhl JP (1996b) Antigen sampling across epithelial barriers and induction of mucosal immune responses. Annu Rev Immunol 14:275–300

Ngo VN, Cornall RJ, Cyster JG (2001) Splenic T zone development is B cell dependent. J Exp Med 194:1649–1660

Odenwald MA, Turner JR (2013) Intestinal permeability defects: is it time to treat? Clin Gastroenterol Hepatol Off Clin Pract J Am Gastroenterol Assoc. 11:1075–1083

Olivier BJ, Cailotto C, van der Vliet J, Knippenberg M, Greuter MJ, Hilbers FW et al (2016) Vagal innervation is required for the formation of tertiary lymphoid tissue in colitis. Eur J Immunol 46:2467–2480

Pabst O, Herbrand H, Friedrichsen M, Velaga S, Dorsch M, Berhardt G et al (2006) Adaptation of solitary intestinal lymphoid tissue in response to microbiota and chemokine receptor CCR7 signaling. J Immunol Baltim Md 1950 177:6824–6832

Parnell EA, Walch EM, Lo DD (2017) Inducible colonic M cells are dependent on TNFR2 but not Ltβr, identifying distinct signalling requirements for constitutive versus inducible M cells. J Crohns Colitis 11:751–760

Patel JR, Williams JL, Muccigrosso MM, Liu L, Sun T, Rubin JB et al (2012) Astrocyte TNFR2 is required for CXCL12-mediated regulation of oligodendrocyte progenitor proliferation and differentiation within the adult CNS. Acta Neuropathol (Berl) 124:847–860

Peduto L, Dulauroy S, Lochner M, Späth GF, Morales MA, Cumano A et al (2009) Inflammation recapitulates the ontogeny of lymphoid stromal cells. J Immunol Baltim Md 1950 182:5789–5799

Peters A, Pitcher LA, Sullivan JM, Mitsdoerffer M, Acton SE, Franz B et al (2011) Th17 cells induce ectopic lymphoid follicles in central nervous system tissue inflammation. Immunity 35:986–996

Picarella DE, Kratz A, Li CB, Ruddle NH, Flavell RA (1993) Transgenic tumor necrosis factor (TNF)-alpha production in pancreatic islets leads to insulitis, not diabetes. Distinct patterns of inflammation in TNF-alpha and TNF-beta transgenic mice. J Immunol Baltim Md 1950 150:4136–4150

Pikor NB, Prat A, Bar-Or A, Gommerman JL (2015) Meningeal tertiary lymphoid tissues and multiple sclerosis: a gathering place for diverse types of immune cells during CNS autoimmunity. Front Immunol 6:657

Poole JA, Wyatt TA, Kielian T, Oldenburg P, Gleason AM, Bauer A et al (2011) Toll-like receptor 2 regulates organic dust-induced airway inflammation. Am J Respir Cell Mol Biol 45:711–719

Poole JA, Gleason AM, Bauer C, West WW, Alexis N, van Rooijen N et al (2012) CD11c(+)/CD11b(+) cells are critical for organic dust-elicited murine lung inflammation. Am J Respir Cell Mol Biol 47:652–659

Poole JA, Mikuls TR, Duryee MJ, Warren KJ, Wyatt TA, Nelson AJ, et al (2017) A role for B cells in organic dust induced lung inflammation. Respir Res [Internet] [cited 19 Apr 2018];18. Available from: https://www.ncbi.nlm.nih.gov/pmc/articles/PMC5741951/

Randolph GJ, Bala S, Rahier J-F, Johnson MW, Wang PL, Nalbantoglu Ilk et al (2016) Lymphoid aggregates remodel lymphatic collecting vessels that serve mesenteric lymph nodes in Crohn disease. Am J Pathol 186:3066–3073

Rangel-Moreno J, Hartson L, Navarro C, Gaxiola M, Selman M, Randall TD (2006) Inducible bronchus-associated lymphoid tissue (iBALT) in patients with pulmonary complications of rheumatoid arthritis. J Clin Invest 116:3183–3194

Ray A, Basu S, Williams CB, Salzman NH, Dittel BN (2012) A novel IL-10-independent regulatory role for B cells in suppressing autoimmunity by maintenance of regulatory T cells via GITR ligand. J Immunol Baltim Md 1950 188:3188–3198

Rehal S, von der Weid P-Y (2017) TNFΔARE mice display abnormal lymphatics and develop tertiary lymphoid organs in the mesentery. Am J Pathol 187:798–807

Sansbury BE, Spite M (2016) Resolution of acute inflammation and the role of resolvins in immunity, thrombosis, and vascular biology. Circ Res 119:113–130

Schrama D, thor Straten P, Fischer WH, McLellan AD, Bröcker EB, Reisfeld RA et al (2001) Targeting of lymphotoxin-alpha to the tumor elicits an efficient immune response associated with induction of peripheral lymphoid-like tissue. Immunity 14:111–121

Schröder AE, Greiner A, Seyfert C, Berek C (1996) Differentiation of B cells in the nonlymphoid tissue of the synovial membrane of patients with rheumatoid arthritis. Proc Natl Acad Sci U S A 93:221–225

Serhan CN, Levy BD (2018) Resolvins in inflammation: emergence of the pro-resolving superfamily of mediators. J Clin Invest 128:2657–2669

Shen P, Fillatreau S (2015) Antibody-independent functions of B cells: a focus on cytokines. Nat Rev Immunol 15:441–451

Shi K, Hayashida K, Kaneko M, Hashimoto J, Tomita T, Lipsky PE et al (2001) Lymphoid chemokine B cell-attracting chemokine-1 (CXCL13) is expressed in germinal center of ectopic lymphoid follicles within the synovium of chronic arthritis patients. J Immunol Baltim Md 1950 166:650–655

Shipman WD, Dasoveanu DC, Lu TT (2017) Tertiary lymphoid organs in systemic autoimmune diseases: pathogenic or protective? F1000Research 6:196

Spalinger MR, McCole DF, Rogler G, Scharl M (2016) Protein tyrosine phosphatase non-receptor type 2 and inflammatory bowel disease. World J Gastroenterol 22:1034–1044

Spalinger MR, Manzini R, Hering L, Riggs JB, Gottier C, Lang S et al (2018) PTPN2 regulates inflammasome activation and controls onset of intestinal inflammation and colon cancer. Cell Rep 22:1835–1848

Srikakulapu P, Hu D, Yin C, Mohanta SK, Bontha SV, Peng L et al (2016) Artery tertiary lymphoid organs control multilayered territorialized atherosclerosis B-cell responses in aged ApoE-/- mice. Arterioscler Thromb Vasc Biol 36:1174–1185

Stalder AK, Carson MJ, Pagenstecher A, Asensio VC, Kincaid C, Benedict M et al (1998) Late-onset chronic inflammatory encephalopathy in immune-competent and severe combined immune-deficient (SCID) mice with astrocyte-targeted expression of tumor necrosis factor. Am J Pathol 153:767–783

Steere AC, Duray PH, Butcher EC (1988) Spirochetal antigens and lymphoid cell surface markers in Lyme synovitis. Comparison with rheumatoid synovium and tonsillar lymphoid tissue. Arthritis Rheum 31:487–495

Takemura S, Braun A, Crowson C, Kurtin PJ, Cofield RH, O'Fallon WM et al (2001) Lymphoid neogenesis in rheumatoid synovitis. J Immunol Baltim Md 1950 167:1072–1080

Tango M, Suzuki E, Gejyo F, Ushiki T (2000) The presence of specialized epithelial cells on the bronchus-associated lymphoid tissue (BALT) in the mouse. Arch Histol Cytol 63:81–89

Tedder TF (2015) B10 cells: a functionally defined regulatory B cell subset. J Immunol Baltim Md 1950 194:1395–1401

Thurlings RM, Wijbrandts CA, Mebius RE, Cantaert T, Dinant HJ, van der Pouw-Kraan TCTM et al (2008) Synovial lymphoid neogenesis does not define a specific clinical rheumatoid arthritis phenotype. Arthritis Rheum 58:1582–1589

Tsuji M, Suzuki K, Kitamura H, Maruya M, Kinoshita K, Ivanov II et al (2008) Requirement for lymphoid tissue-inducer cells in isolated follicle formation and T cell-independent immunoglobulin A generation in the gut. Immunity 29:261–271

Van Itallie CM, Anderson JM (2014) Architecture of tight junctions and principles of molecular composition. Semin Cell Dev Biol 36:157–165

Van Itallie CM, Holmes J, Bridges A, Gookin JL, Coccaro MR, Proctor W et al (2008) The density of small tight junction pores varies among cell types and is increased by expression of claudin-2. J Cell Sci 121:298–305

Wengner AM, Höpken UE, Petrow PK, Hartmann S, Schurigt U, Bräuer R et al (2007) CXCR5- and CCR7-dependent lymphoid neogenesis in a murine model of chronic antigen-induced arthritis. Arthritis Rheum 56:3271–3283

Wood MB, Rios D, Williams IR (2016) TNF-α augments RANKL-dependent intestinal M cell differentiation in enteroid cultures. Am J Physiol Cell Physiol 311:C498–C507

Wu HY, Nguyen HH, Russell MW (1997) Nasal lymphoid tissue (NALT) as a mucosal immune inductive site. Scand J Immunol 46:506–513

Wu Q, Salomon B, Chen M, Wang Y, Hoffman LM, Bluestone JA et al (2001) Reversal of spontaneous autoimmune insulitis in nonobese diabetic mice by soluble lymphotoxin receptor. J Exp Med 193:1327–1332

Yin C, Mohanta SK, Srikakulapu P, Weber C, Habenicht AJR (2016) Artery tertiary lymphoid organs: powerhouses of atherosclerosis immunity. Front Immunol 7:387

Yu P, Lee Y, Liu W, Chin RK, Wang J, Wang Y et al (2004) Priming of naive T cells inside tumors leads to eradication of established tumors. Nat Immunol 5:141–149

Zuercher AW, Coffin SE, Thurnheer MC, Fundova P, Cebra JJ (2002) Nasal-associated lymphoid tissue is a mucosal inductive site for virus-specific humoral and cellular immune responses. J Immunol Baltim Md 1950 168:1796–1803

Role of Lymphoid Structure in Skin Immunity

Gyohei Egawa and Kenji Kabashima

Contents

Abstract The skin is the outermost organ of the body and is exposed to many kinds of external pathogens. To manage this, the skin contains multiple types of immune cells. To achieve sufficient induction of cutaneous adaptive immune responses, the antigen presentation/recognition in the skin is an essential process. Recent studies have expanded our knowledge of how T cells survey their cognate antigens in the skin. In addition, the formation of a lymphoid cluster, named inducible skin-associated lymphoid tissue (iSALT), has been reported during skin inflammation. Although iSALT may not be classified as a typical tertiary lymphoid

G. Egawa (✉) · K. Kabashima
Department of Dermatology, Kyoto University Graduate School of Medicine, Kyoto, Japan
e-mail: gyohei@kuhp.kyoto-u.ac.jp

Current Topics in Microbiology and Immunology (2020) 426: 65–82
https://doi.org/10.1007/82_2020_206

Published Online: 09 May 2020

organ, it provides specific antigen presentation sites in the skin. In this article, we provide an overview of the antigen presentation mechanism in the skin, with a focus on the development of iSALT and its function.

List of Abbreviations

APCs	Antigen-presenting cells
cDCs	Conventional dendritic cells
CHS	Contact hypersensitivity
DCs	Dendritic cells
FRC	Follicular reticular cell
iSALT	Inducible skin-associated lymphoid tissue
LCs	Langerhans cells
LNs	Lymph nodes
MALT	Mucosa-associated lymphoid tissue
PCVs	Post-capillary venules
pDCs	Plasmacytoid DCs
PVMs	Perivascular macrophages
SALT	Skin-associated lymphoid tissue
SLO	Secondary lymphoid organs
TLO	Tertiary lymphoid organ
Tregs	Regulatory T cells
T_{RM}	Resident memory T cells

1 Introduction

The skin is the outermost organ of the body and is exposed to many kinds of external insults. To maintain the homeostasis of such an irritable tissue, the skin is equipped with two types of barrier systems: a physical barrier and an immunological barrier (Egawa and Kabashima 2016).

The physiological barrier of the skin is solid compared with that in mucous membranes, such as in tracheal and gastrointestinal tracts, because the skin needs to block the evaporation of internal water. It consists of the stratum corneum (a cornified layer of dead keratinocytes) and the tight junctions in the stratum granulosum, which are covered with the sebum secreted from sebaceous glands (Fig. 1). These barriers are robust and block molecules with a molecular weight greater than 1000; therefore, no pathogenic microorganism can invade the body through the intact physical skin barrier. In reality, however, there are many open seams in the physical barrier; the skin easily sustains small traumas and the skin appendages, such as hair follicles and sweat pores, lack the stratum corneum and can harbor

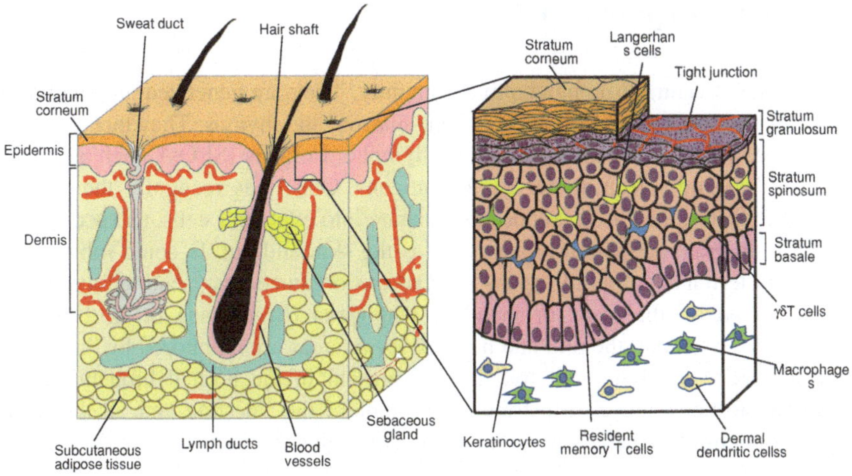

Fig. 1 The physical and immunological barrier of the skin. In the epidermis, tight junctions are formed underneath the stratum corneum (in the stratum granulosum), and three types of immune cells (Langerhans cells, γδ T cells, and resident memory T cells) may reside between keratinocytes

microorganisms, known as the skin microbiome (Kabashima et al. 2019) (Fig. 1; left). The pathogenic microorganisms/molecules often invade the body through such "security holes" in the skin. To manage these invaders, the immune system keeps the skin under constant surveillance.

The immunological barrier of the skin consists of multiple types of immune cells (Fig. 1; right). Skin-resident innate immune cells, such as epidermal Langerhans cells (LCs), dermal dendritic cells (DCs), and macrophages, monitor the invasion of foreign antigens. Once they recognize the antigens, they produce inflammatory cytokines to cause inflammation. Upon inflammation, many immune cells, such as neutrophils, monocytes, and T cells, are recruited to the skin. Neutrophils and monocytes provide a quick but unspecific immune response, whereas T cells provide a delayed but antigen-specific immune reaction. For the latter reaction, antigen presentation to T cells in the skin is an essential process. To facilitate this, lymphoid structures can be formed in the skin in some pathogenic conditions (discussed later).

In this article, we will discuss the antigen presentation mechanism in the skin and focus on the function of a specific immunological unit, called inducible skin-associated lymphoid tissue (iSALT), which plays an essential role in the induction of cutaneous adaptive immunity (Natsuaki et al. 2014). We also review the formation of lymphoid clusters in the human skin in several pathogenic conditions.

2 The Concept of SALT

T cells play a central role in adaptive immunity. They are generated and mature in the primary lymphoid organs such as bone marrow and thymus. They then circulate in the blood and travel to the secondary lymphoid organs (SLOs), such as lymph nodes (LNs) and spleen, to survey for antigens. Once they recognize the cognate antigen in SLOs, they are expanded and mature into effector T cells, and acquire the nature to migrate peripheral organs such as the skin and gut. In some submucosal areas, the tertiary lymphoid organs (TLOs) are organized as "sentinel" lymphoid tissues to be on the alert for invading pathogens (Brandtzaeg et al. 1999). In humans, for example, the oral and nasal pharynx areas are monitored by the tonsils and adenoids, and lymphoid follicles are present in the normal bronchi. Single lymphoid follicles are also distributed throughout the intestine and, in the distal ileum, lymphoid follicles are grouped in large clusters termed Peyer's patches. These tissues are known as mucosa-associated lymphoid tissue (MALT) and serve as antigen presentation sites in peripheral organs. As for skin, however, no lymphoid structures were reported until recently.

Around 1980, cutaneous immunologists elucidated some key findings; (1) LCs are bone marrow-derived and capable of antigen presentation, (2) a fraction of T cells display high skin tropism, and (3) epidermal cells markedly affect T cell maturation by producing multiple cytokines and chemokines (Stingl et al. 1978; Rubenfeld et al. 1981). Based on these findings, researchers proposed that lymphoid tissues analogous to MALT in submucosal areas may exist within the skin. They offered the term SALT (skin-associated lymphoid tissue) for these putative skin-associated tissues (Streilein 1978, 1983, 1985; Egawa and Kabashima 2011). As lymphoid structure was not found in the skin at that time, SALT was a conceptual tissue; however, this hypothesis proposes that the skin is not merely a physical barrier but also an essential component of the immune system, and that the antigen presentation in the skin is an important step in elicitation of acquired skin immune responses.

3 Antigen Presentation Mechanism in the Skin

After the development of the concept of SALT, focus was placed on the antigen presentation mechanism in the skin. As the majority of skin-infiltrating T cells are memory-phenotype, and naïve T cells and B cells are almost absent in the skin (Clark et al. 2006), the antigen presentation process in the skin should be substantially different from that in the LNs and MALTs. Recent studies with transgenic animals in combination with intravital imaging techniques have extensively expanded our knowledge of the biology of cutaneous antigen-presenting cells (APCs) and skin-resident/homing T cells. In this section, we will overview how cells in the skin induce inflammation against pathogen invasion and how T-cell-mediated adaptive immunity is initiated in the skin.

3.1 Keratinocytes as an Initiator of Skin Inflammation

When keratinocytes are injured or recognize pathogens mainly through pattern recognition receptors, such as Toll-like receptors, they produce several kinds of proinflammatory cytokines, including tumor necrosis factor (TNF)-α, interleukin (IL)-1α/β, IL-33, and thymic stromal lymphopoietin (TSLP) (Ansel et al. 1990; Leyva-Castillo et al. 2013; Meephansan et al. 2013). These cytokines induce the expression of E- and P-selectin and intercellular adhesion molecule 1 (ICAM-1) on vascular endothelial cells that promote the extravasation of blood-circulating lymphocytes (Brinkman et al. 2013). Phagocytes, such as neutrophils and monocytes, appear in the skin as the primary treatment, and a small number of T cells arrive slightly later to scan for their cognate antigens. Proinflammatory cytokines produced by keratinocytes also activate skin-resident immune cells, including LCs and DCs. For example, TSLP from keratinocytes promotes the migration of LCs to the skin-draining LNs and enhances the induction of Th2-type immune responses (Nakajima et al. 2012). These functions of keratinocytes as an initiator of skin inflammation suggest that keratinocytes are not only a producer of the physical barrier, but also an inducer of the immunological barrier.

3.2 Cutaneous Antigen-Presenting Cells

The antigens that breach the physical barrier of the skin are then captured by the second-line immunological barrier, cutaneous DCs. DCs are a diverse family of cells that play an essential role in linking the innate and adaptive immune systems (Dress et al. 2018). In general, four cell types are classified into the DC family in the skin: epidermal LCs, conventional DCs (cDCs), plasmacytoid DCs (pDCs), and monocyte-derived DCs. cDCs can be further subdivided into the cDC1 and cDC2 subsets on the basis of subset-specific gene expression profiles, their dependence on different transcription factors, and unique subset functions (Schlitzer et al. 2015; Guilliams et al. 2016).

Epidermal LCs were previously thought to play a major role in antigen presentation not only in the LNs but also in the skin (Toews et al. 1980; Grabbe and Schwarz 1998). However, novel depletion systems of LCs (and dermal DCs) have challenged this classic LC paradigm (Bennett et al. 2005; Kissenpfennig et al. 2005). When LCs were selectively depleted, the contact hypersensitivity (CHS) response, a classical mouse model for type IV hypersensitivity, was not (Kissenpfennig et al. 2005; Bursch et al. 2007; Wang et al. 2008) or only marginally attenuated (Bennett et al. 2005). On the other hand, the depletion of LCs caused impaired $CD8^+$ T cell activation in the skin in a graft-versus-host disease model (Bennett et al. 2011). These reports suggest that the role of LCs as cutaneous APCs differs depending on the pathogenic conditions.

The antigen-presenting role of cDCs is also controversial. Several studies found that the depletion of both LCs and cDC1 markedly reduced the CHS response (Bursch et al. 2007; Wang et al. 2008), suggesting that cDC1, but not LCs, plays an important role during CHS. However, another study reported that Batf3-deficient mice, which lack cDC1 in the skin, exhibited a normal CHS phenotype (Edelson et al. 2010), suggesting compensation of its function by other DCs. We recently demonstrated that when all subsets of cutaneous DCs were depleted, the elicitation of CHS was abrogated, whereas selective depletion of LCs or cDC1 did not attenuate the elicitation of CHS (Natsuaki et al. 2014). These results suggest the dominant role of cDC2 in T cell activation in the skin, at least during CHS.

In addition to DCs, several other skin-resident cells have been proposed to function in antigen presentation in the skin. Mast cells and basophils participate in antigen presentation and promote Th2-type inflammatory responses by acquiring MHC class II molecules from DC, termed "trogocytosis" (Dudeck et al. 2017; Miyake et al. 2017). In addition, the deletion of MHC class I molecule on radio-resistant cells impaired the activation of $CD8^+$ T cells in CHS (Ono et al. 2018), suggesting the possible involvement of vascular endothelial cells (Kish et al. 2011) and keratinocytes (Gaspari and Katz 1988; Kim et al. 2009) as APCs in the skin, although in vivo evidence is lacking. As co-stimulatory molecules are not required for the activation of effector T cells (Krummel et al. 1999), non-professional APCs in the skin may carry out the APC function in a context-dependent manner.

3.3 T Cell Recruitment to the Skin

To mediate their effector functions, most lymphocytes, except for B cells, need to home into the affected peripheral tissues before their immunomodulatory functions can be initiated. Upon inflammation, effector T cells are recruited to the skin with limited antigen dependency (Honda et al. 2014). T cells migrate into the inflamed tissues by scanning vascular endothelial cells displaying molecular signatures required for precise spatial homing. After activation and differentiation within the secondary lymphoid organs, the effector T cells downregulate lymph node homing molecules like CD62L and CCR7 and upregulate molecules specific for their homing into peripheral tissues (Masopust and Schenkel 2013).

Recent studies suggest that effector T cells express homing molecules specific for organs from where the antigenic insult originated. T cells primed in the skin-draining lymph nodes upregulate E- and P-selectin ligands (Tietz et al. 1998; Hirata et al. 2002), CCR4 and/or CCR10, which are required for skin homing (Soler et al. 2003; Masopust and Schenkel 2013). Analogous to vitamin A-mediated upregulation of gut homing molecules (Mora et al. 2008), DCs can process vitamin D present in the skin to its active metabolite 1,25-dihydroxyvitamin D3, which facilitates the induction of CCR10, and concomitant downregulation of $\alpha4\beta7$ and CCR9 (Sigmundsdottir et al. 2007; Masopust and Schenkel 2013). Expression of

CCL17 by skin venules and CCL27 by epidermal keratinocytes helps in homing $CCR4^+$ and $CCR10^+$ T cells to the skin, respectively (Reiss et al. 2001; Homey et al. 2002; Soler et al. 2003). These studies provide mechanistic insight into how the site of priming can imprint homing molecules on effector T cells.

3.4 Antigen Survey by T Cells in the Skin

After homing into the inflamed tissue, T cells have to search for rare antigen-bearing target cells before effector functions can be initiated. Intravital imaging studies have revealed that effector T cells entering the interstitium of the skin exhibit a polarized cell shape, characterized by the formation of a leading edge and a uropod, and migrate at high velocities (Matheu et al. 2008; Egawa et al. 2011; Honda et al. 2014). During interstitial migration, T cells constantly integrate molecular cues provided by surrounding cells. Unlike neutrophils and DCs, T cells are less dependent on chemoattractant gradients. During interstitial migration, they demonstrate a "stop and go" behavior reminiscent of naïve T cells in the LNs (Miller et al. 2002; Kawakami et al. 2005; Munoz et al. 2014; Weninger et al. 2014). This behavior may provide an effective strategy for screening large regions of the tissue.

Once they encounter their cognate antigens, effector T cells initiate stable contact with APCs and are activated to produce inflammatory cytokines. Upon activation, effector $CD4^+$ T cells can produce large-scale cytokine gradients. For example, in the *Leishmania major* infection model, effector $CD4^+$ T cells interact with infected APCs, resulting in migratory arrest and T-cell-mediated production of interferon-γ (IFN-γ), leading to the generation of IFN-γ gradients up to 80 μm away from the T cell-APC interaction site (Muller et al. 2012). On the other hand, cytotoxic effector $CD8^+$ T cells induce target cell apoptosis by different means, including the release of cytokines and cytotoxic mediators such as perforin and granzymes. Although the migratory behavior of effector $CD8^+$ T cell populations within inflamed tissues is best described as random migration, several studies have observed T cell navigation along the extracellular matrix and other anatomical structures. Intravital imaging of subcutaneous tumors has revealed that $CD8^+$ T cells migrate along dermal collagen fibers and blood vessels (Mrass et al. 2006; Boissonnas et al. 2007).

Of note, after infiltration into the skin, a part of the T cells returns back to the draining LNs via afferent lymphatics. Classic lymph recirculation studies in sheep have demonstrated that effector/memory $CD4^+$ T cells comprise a major portion of lymphocytes in afferent lymph (Mackay et al. 1988, 1990). T cell egress from the skin is dependent on CCR7 (Bromley et al. 2005; Debes et al. 2005), similarly to other leukocytes such as DCs or neutrophils (Randolph 2001; Ng et al. 2011). Endothelial cells of the afferent lymph vessels constitutively express the CCR7 ligand CCL21 (Debes et al. 2005). Moreover, a recent study employing photo-convertible Kaede transgenic mice, which serves an in vivo cell-tracking system, reported that the cell migration from peripheral tissues to the draining LNs was markedly increased after inflammation (Tomura et al. 2010). Importantly, approximately half of the $CD4^+$ T cells that return back from the inflamed site are regulatory T cells (Tregs). These observations suggest that Tregs preferentially infiltrate or are induced in peripheral tissue during the resolution phase of inflammation.

3.5 Antigen Recognition by Resident Memory T Cells in the Skin

Although T cells have a dynamic migratory nature, a number of studies suggested that some of the tissue-infiltrating effector T cells never return to the circulation (Shin and Iwasaki 2013; Mueller et al. 2014). This non-circulating memory T cell subset is called tissue-resident memory T cells (T_{RM}). T_{RM} have been identified in several non-lymphoid tissues, such as the skin, gut, lung, brain, and female reproductive tract (Gebhardt et al. 2009; Masopust et al. 2010; Wakim et al. 2010), although the longevity of T_{RM} greatly differs among tissues. For example, skin T_{RM} in mice persist for over a year (Mackay et al. 2012), whereas lung T_{RM} are maintained for a few months (Wu et al. 2014).

Of note, the distribution and migration of $CD4^+$ T_{RM} and $CD8^+$ T_{RM} differ significantly in the skin. Previous studies with herpes simplex virus (HSV) skin infection elucidated that after pathogen clearance, antigen-specific $CD4^+$ T_{RM} redistribute within the dermis, whereas $CD8^+$ T_{RM} localize to the epidermis (Gebhardt et al. 2011; Mackay et al. 2012) (Fig. 2). $CD4^+$ T_{RM} have amoeboid morphology and actively migrate throughout the dermis, whereas $CD8^+$ T_{RM} exhibit a dendritic morphology and are almost sessile. However, during vaccinia virus skin infections, $CD8^+$ T_{RM} were found in the dermis as well as the epidermis, and these $CD8^+$ T_{RM} were found not only in the infected site, but also in the non-infected regions within the tissue (Jiang et al. 2012).

There is evidence suggesting that rapid control of infection at peripheral sites requires the presence of T_{RM}. Upon viral re-infection, T_{RM} respond to antigen and produce proinflammatory cytokines, such as IFN-γ, within hours, whereas circulating memory T cells re-enter the infected site within 2 days but do not produce IFN-γ until 5 days after the challenge (Iijima and Iwasaki 2014). T_{RM}-derived cytokines activate local innate immunity by driving antiviral/antibacterial genes, DC maturation, NK cells activation, and VCAM-1 expression on the blood endothelium (Ariotti et al. 2014; Schenkel et al. 2014). Therefore, T_{RM} function as antigen-specific sensors and provide robust site-specific immunity. The presence of T_{RM} in epidermis should be important in maintaining immunity against HSV and human immunodeficiency virus (HIV), as both of these viruses commonly begin as local infections in the genital tract in a limited population of infected cells. In such situations, T_{RM}-mediated rapid initiation of the antiviral state at the site of entry is essential for preventing re-infections.

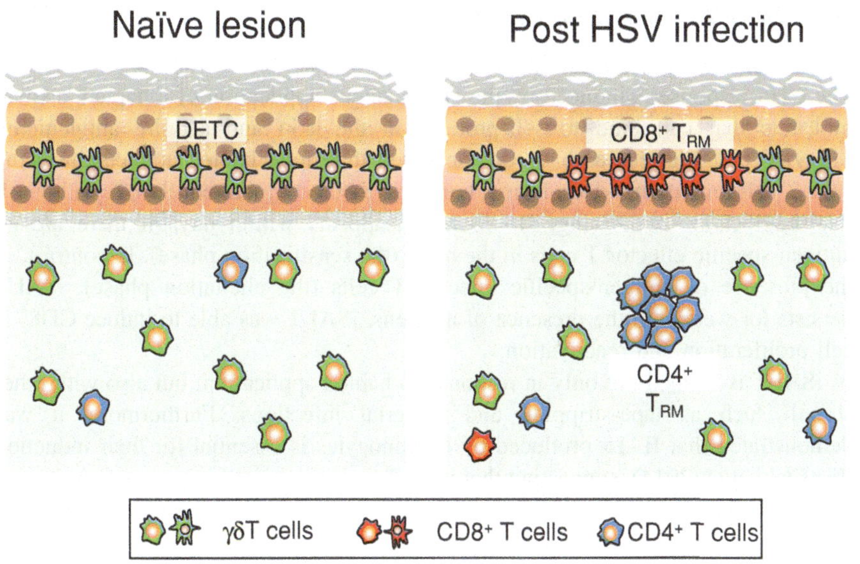

Fig. 2 A schematic representation of the skin depicting the presence of different leukocyte subsets before and after Herpes simplex virus (HSV) infection. In naïve skin, epidermis is populated with sessile Langerhans cells and dendritic epithelial T cells (DETCs), whereas the dermis is populated with dermal DCs, $CD4^+$ T cells, dermal γδ T cells, and a few $CD8^+$ T cells. After the resolution of HSV infection, a subset of antigen-specific $CD8^+$ T_{RM} cells resides in the epidermis, whereas antigen-specific $CD4^+$ T_{RM} clusters reside in the dermis

4 The Concept of iSALT

In contrast to mucosal boundary tissues, such as gut and bronchi, in which TLOs can be observed even in the non-pathogenic condition, no lymphocytic cluster is observed in normal skin. This is probably due to the anatomical difference between the skin and mucosal boundary tissues; external antigens hardly reach the skin through the thick physical barrier, the stratum corneum. Therefore, it has been generally considered that skin-infiltrating T cells randomly migrate in the skin to scan for cognate antigens, and that individual T cell-APC interaction is important for T cell activation in the skin.

Recently, however, the formation of a leukocyte-clustering structure was found during skin inflammation (Natsuaki et al. 2014). Unlike MALT, these leukocyte clusters are not found in the steady state and are inducible during the development of acquired immune response. Thus, this cluster was named inducible skin-associated lymphoid tissue (iSALT), similar to inducible bronchus-associated lymphoid tissue (iBALT) in the lung (Moyron-Quiroz et al. 2004). In this section, we will overview the mechanism of iSALT formation, especially during the CHS response.

4.1 iSALT Formation in CHS

The CHS response is induced by the epicutaneous application of a hapten, a small molecule that can bind to self-proteins in the skin and acquire antigenicity. Intravital imaging studies have demonstrated that upon hapten application to the skin, T cells and dermal DCs form a lymphocytic cluster, termed iSALT, within 6 h (Natsuaki et al. 2014) (Fig. 3). iSALT disappears within days if there are no antigen-specific effector T cells in the body (the sensitization phase). In contrast, in the presence of antigen-specific effector T cells (the elicitation phase), iSALT persists for weeks. In the presence of antigens, iSALT was able to induce $CD8^+$ T cell proliferation and reactivation.

iSALT is formed not only in response to hapten application, but also with other stimuli such as tape-stripping and bacterial infections. Furthermore, it was demonstrated that IL-1α produced by keratinocytes is essential for their induction (Natsuaki et al. 2014), suggesting that iSALT formation itself is an innate response.

4.2 iSALT Formation Around Post-capillary Venules

Dermal blood vessels can be divided into four different parts with distinct functions: arteries, capillaries, post-capillary venules (PCVs), and venules. Among them, PCVs have a unique property that is particularly important during inflammation. Adjacent blood endothelial cells are sealed with tight junctions and adherence junctions, and limit the passage of plasma proteins larger than 70 kDa (Egawa et al. 2013). Importantly, this vascular permeability is variable only at PCVs. Upon inflammation, hyper-permeability is induced on PCVs, leading to extravasation of albumin (70 kDa) and immunoglobulins (150 kDa) into the dermal interstitium, suggesting that PCVs are specific portal sites for humoral immunity into the skin under inflammatory conditions (Fig. 4).

Intravital studies also demonstrated that PCVs are specific portal sites for cellular immunity. In PCVs, blood endothelial cells are surrounded by pericytes and

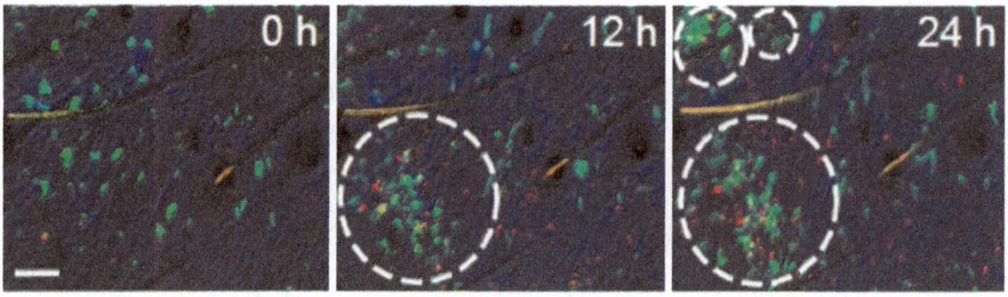

Fig. 3 Intravital images of iSALT formation in a murine CHS model. Dermal DCs (green) and T cell (red) form clusters in the dermis (white circles) within hours. Scale bar = 100 μm

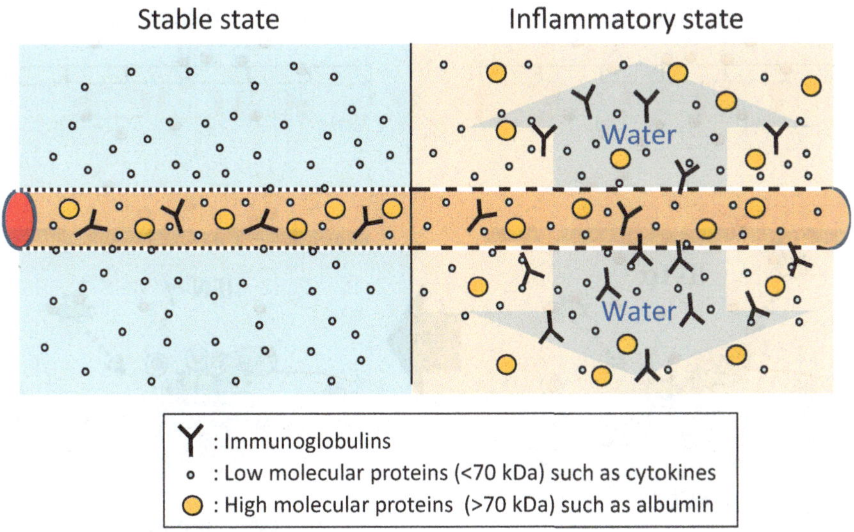

Fig. 4 A scheme of hyper-permeabilization in PCVs. In the steady state, only plasma contents having a molecular size less than 70 kDa can extravasate, whereas in the inflammatory state, large plasma contents (>70 kDa), such as albumin and immunoglobulins, freely pass through the blood vessel walls

macrophages, and mast cells are located nearby. Previous studies using a mouse *Staphylococcus aureus* infection model revealed that these cellular units around PCVs are important for neutrophil extravasation in the dermis (Abtin et al. 2014). As for T cells, intravital imaging studies and immunohistochemical analyses revealed that most iSALT was formed around PCVs (Honda and Kabashima 2016; Kogame et al. 2017) (Fig. 5) and in the absence of iSALT formation, T cell infiltration to the skin is impaired. These findings suggest that cellular units around PCVs play an import role in lymphocyte recruitment into the skim. In particular, recent studies revealed indispensable roles of perivascular macrophages (PVMs) in this process.

4.3 *Indispensable Role of Perivascular Macrophages in iSALT Formation*

An intravital imaging study using DPE-GFP mice revealed that approximately 40% of venules are surrounded by PVMs (Abtin et al. 2014). Upon *S. aureus* infection, PVMs highly express the neutrophil-attracting chemokines Cxcl1 and Cxcl2, and neutrophils extravasate in the vicinity of PVMs. PVMs are depleted when they are exposed to *S. aureus*-producing exotoxin, a-hemolysin, and under this condition, neutrophil recruitment to the skin is significantly suppressed (Abtin et al. 2014).

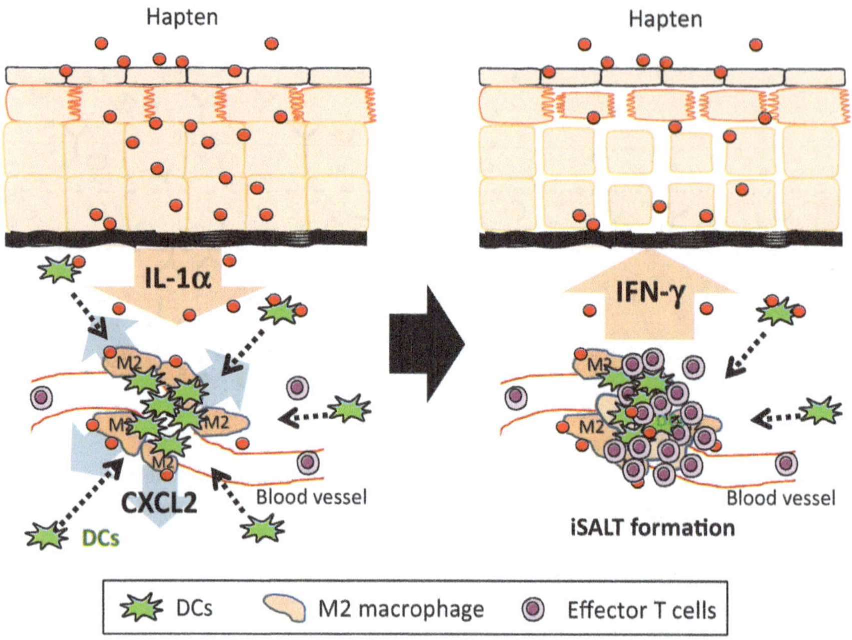

Fig. 5 A schema of iSALT formation during CHS. Keratinocytes contact with hapten induces the release of IL-1 in the skin, which activates perivascular macrophages that subsequently attract dermal DCs via CXCR2 to form clusters. In the absence of antigen-specific memory T cells, DC clustering is a transient event, and hapten-carrying DCs migrate to the skin-draining LNs. In the presence of antigen-specific effector T cells, T cells are activated and skin inflammation is promptly induced

PCVs also play an essential role during iSALT formation and T cell activation in the skin (Natsuaki et al. 2014). Cell-type specific depletion studies found that mast cells, T cells, B cells, and basophils were dispensable for iSALT formation by hapten application, but when macrophages were depleted, iSALT formation was abrogated. Further analysis revealed that IL-1α produced by keratinocytes upon external insults stimulates M2-like macrophages around PCVs, which then produce Cxcl2 and recruit dermal DCs to the cluster to form iSALT. Leukotriene B4, a lipid mediator, also mediates the cluster formation by promoting DC migration (Sawada et al. 2015). Subsequently, effector T cells accumulated in the cluster are presented antigens by dermal DCs and initiate proliferation and activation. In the cluster, both dermal DC subsets (Kashem et al. 2017), i.e., $CD103^+$ cDC1 and $CD11b^+$ cDC2, are detected (Okada et al. 2016). It is currently unclear, however, which dermal DC subsets mediate antigen presentation in the cluster (Ono et al. 2018). Each DC subset in the cluster may function in a compensatory manner like DCs in the sensitization phase (Honda et al. 2010). The blockade of CXCL2 and IL-1 receptor signaling impairs iSALT formation and effector T cell activation, suggesting that iSALT is an essential structure for antigen presentation in the skin.

5 iSALT Formation in Humans

It remains unclear whether iSALT formation and functions in T cell activation in human skin are analogous to those in mice. In many inflammatory skin diseases, including eczema, psoriasis, and drug eruptions, perivascular leukocyte infiltrations are frequently observed by histological examination. In allergic contact dermatitis in humans, T cell-DC clusters are found in the dermis and are accompanied by vesicle formation in the epidermis, a marker of eczema, suggesting T cell activation and subsequent cytokine production occur above the cluster (Natsuaki et al. 2014).

In psoriasis patients, the existence of high endothelial venules (HEV), a characteristic structure of TLOs, has been reported (Lowe et al. 1995), and clusters of DC-LAMP^{+} DCs and T cells have been detected in the dermis (Zaba et al. 2007) with abundant expression of CCL19, a ligand for CR7, and CCL20 (Mitsui et al. 2012; Kim et al. 2014). CCL20 is implicated in the formation and function of MALT via the chemoattraction of CCR6^{+} lymphocytes and DCs. This cluster observed in psoriatic lesions disappears after treatment with TNF-α inhibitors (Zaba et al. 2007), suggesting the fundamental role of TNF-α in the maintenance of the cluster.

iSALT-like structures were also found in the skin lesions of secondary syphilis infection (Kogame et al. 2017). Of note, these lymphatic clusters contain CXCL13^{+} cells. CXCL13 is a marker of follicular helper T cells and is an important chemokine responsible for the formation and maintenance of lymphatic clusters with B cells. Indeed, spotty infiltration of B cells is observed in the skin lesions of secondary syphilis infection. In melanoma patients, clusters with TLO features, such as the existence of HEV, T cells, B cells, and mature DCs, have been detected in the extratumoral area with tumor regression or favorable overall survival (Ladanyi et al. 2007; Martinet et al. 2012). Lymphoid follicles are also reported in the lesional skin of cutaneous lupus erythematosus (Arps and Patel 2013) and lymphoproliferative diseases such as Kimura's disease (Kung et al. 1984). Although the functional significance of the leukocyte clusters/lymphoid follicles in the skin remains unclear, these structures may play important roles in the promotion or regulation of disease development.

6 Future Remarks

TLOs play important roles in host protection and the development of pathogenic conditions in non-lymphoid peripheral tissues (Dieu-Nosjean et al. 2014; Pitzalis et al. 2014; Colbeck et al. 2017). TLOs have not yet been clearly defined, but they should fulfill several characteristics: (1) the existence of distinct T and B cell compartments, (2) a follicular reticular cell (FRC) network, (3) peripheral node addressins (PNAd)$^{+}$ HEVs, (4) lymphatic vasculature, and (5) evidence of class switching (Dieu-Nosjean et al. 2014). Most TLOs are not genetically programmed and do not develop postnatally. Lymphotoxin, lymphoid chemokines (CCL19, CCL21, CXCL13), TNF-α, and the receptor activator of nuclear factor kappa-B ligand (RANKL) play essential roles in the development of TLOs.

Based on these structural criteria, iSALT may not be classified as a typical TLO because of the absence of B cells, naïve T cells, HEV, and the FRC network. In addition, the involvement of lymphoid chemokines has not yet been fully clarified in iSALT. From a functional point of view, however, iSALT possesses the key features of a TLO in that it offers efficient sites for effector T cell activation; some researchers thus use the term TLO for iSALT, focusing on this point (Neyt et al. 2012; Colbeck et al. 2017).

Although both MALT and iSALT provide defined sites for antigen presentation within peripheral organs, there should be distinct functional differences between these tissues. MALT contains significant numbers of B cells and forms lymph follicles, whereas virtually all lymphocytes in the iSALT are T cells. MALT contains HEVs and serves as an entry point for naïve T cells, suggesting that it is equipped to provide a field for antigen presentation to naïve T cells and other SLOs. In contrast, HEVs are rare in the skin and most of the T cells recruited to the skin are considered to be antigen-experienced effector T cells. Thus, SALT may act as a peripheral lymphoid tissue to provide a function distinct from other secondary/tertiary lymphoid organs, including MALT.

Although studies on iSALT function have progressed, we lack a comprehensive understanding of the skin in health and diseases. It is unclear which immune cells and/or non-immune cells interact with each other at which immune response time point (e.g., the acute, chronic, and resolution stages) and where in the skin. Considering the fundamental differences between mouse and human skin is another important challenge. Evaluation of these points may lead to a breakthrough in the understanding of the immunological mechanisms of cutaneous immune responses and healthy skin homeostasis.

References

Abtin A et al (2014) Perivascular macrophages mediate neutrophil recruitment during bacterial skin infection. Nat Immunol 15(1):45

Ansel J et al (1990) Cytokine modulation of keratinocyte cytokines. J Invest Dermatol 94(6):s101–s107

Ariotti S et al (2014) T cell memory. Skin-resident memory CD8(+) T cells trigger a state of tissue-wide pathogen alert. Science 346(6205):101–105

Arps DP, Patel RM (2013) Lupus profundus (panniculitis): a potential mimic of subcutaneous panniculitis-like T-cell lymphoma. Arch Pathol Lab Med 137(9):1211–1215

Bennett CL et al (2011) Langerhans cells regulate cutaneous injury by licensing CD8 effector cells recruited to the skin. Blood 117(26):7063–7069

Bennett CL et al (2005) Inducible ablation of mouse Langerhans cells diminishes but fails to abrogate contact hypersensitivity. J Cell Biol 169(4):569–576

Boissonnas A et al (2007) In vivo imaging of cytotoxic T cell infiltration and elimination of a solid tumor. J Exp Med 204(2):345–356

Brandtzaeg P et al (1999) Regional specialization in the mucosal immune system: what happens in the microcompartments? Immunol Today 20(3):141–151

Brinkman CC et al (2013) Peripheral tissue homing receptor control of naive, effector, and memory CD8 T cell localization in lymphoid and non-lymphoid tissues. Front Immunol 4:241

Bromley SK et al (2005) Chemokine receptor CCR7 guides T cell exit from peripheral tissues and entry into afferent lymphatics. Nat Immunol 6(9):895–901

Bursch LS et al (2007) Identification of a novel population of Langerin + dendritic cells. J Exp Med 204(13):3147–3156

Clark RA et al (2006) The vast majority of CLA + T cells are resident in normal skin. J Immunol 176(7):4431–4439

Colbeck EJ et al (2017) Tertiary lymphoid structures in cancer: drivers of antitumor immunity, immunosuppression, or bystander sentinels in disease? Front Immunol 8:1830

Debes GF et al (2005) Chemokine receptor CCR7 required for T lymphocyte exit from peripheral tissues. Nat Immunol 6(9):889–894

Dieu-Nosjean MC et al (2014) Tertiary lymphoid structures in cancer and beyond. Trends Immunol 35(11):571–580

Dress RJ et al (2018) Homeostatic control of dendritic cell numbers and differentiation. Immunol Cell Biol

Dudeck J et al (2017) Mast cells acquire MHCII from dendritic cells during skin inflammation. J Exp Med 214(12):3791–3811

Edelson BT et al (2010) Peripheral CD103 + dendritic cells form a unified subset developmentally related to CD8alpha + conventional dendritic cells. J Exp Med 207(4):823–836

Egawa G et al (2011) In vivo imaging of T-cell motility in the elicitation phase of contact hypersensitivity using two-photon microscopy. J Invest Dermatol 131(4):977–979

Egawa G, Kabashima K (2011) Skin as a peripheral lymphoid organ: revisiting the concept of skin-associated lymphoid tissues. J Invest Dermatol 131(11):2178–2185

Egawa G, Kabashima K (2016) Multifactorial skin barrier deficiency and atopic dermatitis: essential topics to prevent the atopic march. J Allergy Clin Immunol 138(2):350–358. e351

Egawa G et al (2013) Intravital analysis of vascular permeability in mice using two-photon microscopy. Sci Rep 3:1932

Gaspari AA, Katz SI (1988) Induction and functional characterization of class II MHC (Ia) antigens on murine keratinocytes. J Immunol 140(9):2956–2963

Gebhardt T et al (2009) Memory T cells in nonlymphoid tissue that provide enhanced local immunity during infection with herpes simplex virus. Nat Immunol 10(5):524–530

Gebhardt T et al (2011) Different patterns of peripheral migration by memory CD4+ and CD8+ T cells. Nature 477(7363):216–219

Grabbe S, Schwarz T (1998) Immunoregulatory mechanisms involved in elicitation of allergic contact hypersensitivity. Immunol Today 19(1):37–44

Guilliams M et al (2016) Unsupervised high-dimensional analysis aligns dendritic cells across tissues and species. Immunity 45(3):669–684

Hirata T et al (2002) P-, E-, and L-selectin mediate migration of activated CD8+ T lymphocytes into inflamed skin. J Immunol 169(8):4307–4313

Homey B et al (2002) CCL27-CCR10 interactions regulate T cell-mediated skin inflammation. Nat Med 8(2):157–165

Honda T et al (2010) Compensatory role of Langerhans cells and langerin-positive dermal dendritic cells in the sensitization phase of murine contact hypersensitivity. J Allergy Clin Immunol 125(5):1154–1156 e1152

Honda T et al (2014) Tuning of antigen sensitivity by T cell receptor-dependent negative feedback controls T cell effector function in inflamed tissues. Immunity 40(2):235–247

Honda T, Kabashima K (2016) Novel concept of iSALT (inducible skin-associated lymphoid tissue) in the elicitation of allergic contact dermatitis. Proc Jpn Acad Ser B Phys Biol Sci 92 (1):20–28

Iijima N, Iwasaki A (2014) T cell memory. A local macrophage chemokine network sustains protective tissue-resident memory CD4 T cells. Science 346(6205):93–98

Jiang X et al (2012) Skin infection generates non-migratory memory CD8+ T(RM) cells providing global skin immunity. Nature 483(7388):227–231
Kabashima K et al (2019) The immunological anatomy of the skin. Nat Rev Immunol 19(1):19–30
Kashem SW et al (2017) Antigen-presenting cells in the skin. Annu Rev Immunol 35:469–499
Kawakami N et al (2005) Live imaging of effector cell trafficking and autoantigen recognition within the unfolding autoimmune encephalomyelitis lesion. J Exp Med 201(11):1805–1814
Kim BS et al (2009) Keratinocytes function as accessory cells for presentation of endogenous antigen expressed in the epidermis. J Invest Dermatol 129(12):2805–2817
Kim TG et al (2014) Dermal clusters of mature dendritic cells and T cells are associated with the CCL20/CCR6 chemokine system in chronic psoriasis. J Invest Dermatol 134(5):1462–1465
Kish DD et al (2011) Hapten application to the skin induces an inflammatory program directing hapten-primed effector CD8 T cell interaction with hapten-presenting endothelial cells. J Immunol 186(4):2117–2126
Kissenpfennig A et al (2005) Dynamics and function of Langerhans cells in vivo: dermal dendritic cells colonize lymph node areas distinct from slower migrating Langerhans cells. Immunity 22 (5):643–654
Kogame T et al (2017) Possible inducible skin-associated lymphoid tissue (iSALT)-like structures with CXCL13(+) fibroblast-like cells in secondary syphilis. Br J Dermatol 177(6):1737–1739
Krummel MF et al (1999) Differential coupling of second signals for cytotoxicity and proliferation in CD8+ T cell effectors: amplification of the lytic potential by B7. J Immunol 163(6):2999–3006
Kung IT et al (1984) Kimura's disease: a clinico-pathological study of 21 cases and its distinction from angiolymphoid hyperplasia with eosinophilia. Pathology 16(1):39–44
Ladanyi A et al (2007) Density of DC-LAMP(+) mature dendritic cells in combination with activated T lymphocytes infiltrating primary cutaneous melanoma is a strong independent prognostic factor. Cancer Immunol Immunother 56(9):1459–1469
Leyva-Castillo JM et al (2013) TSLP produced by keratinocytes promotes allergen sensitization through skin and thereby triggers atopic march in mice. J Invest Dermatol 133(1):154–163
Lowe PM et al (1995) The endothelium in psoriasis. Br J Dermatol 132(4):497–505
Mackay CR et al (1988) Lymphocyte subsets show marked differences in their distribution between blood and the afferent and efferent lymph of peripheral lymph nodes. J Exp Med 167 (6):1755–1765
Mackay CR et al (1990) Naive and memory T cells show distinct pathways of lymphocyte recirculation. J Exp Med 171(3):801–817
Mackay LK et al (2012) Long-lived epithelial immunity by tissue-resident memory T (TRM) cells in the absence of persisting local antigen presentation. Proc Natl Acad Sci USA 109(18):7037–7042
Martinet L et al (2012) High endothelial venules (HEVs) in human melanoma lesions: Major gateways for tumor-infiltrating lymphocytes. Oncoimmunology 1(6):829–839
Masopust D et al (2010) Dynamic T cell migration program provides resident memory within intestinal epithelium. J Exp Med 207(3):553–564
Masopust D, Schenkel JM (2013) The integration of T cell migration, differentiation and function. Nat Rev Immunol 13(5):309–320
Matheu MP et al (2008) Imaging of effector memory T cells during a delayed-type hypersensitivity reaction and suppression by Kv1.3 channel block. Immunity 29(4):602–614
Meephansan J et al (2013) Expression of IL-33 in the epidermis: the mechanism of induction by IL-17. J Dermatol Sci 71(2):107–114
Miller MJ et al (2002) Two-photon imaging of lymphocyte motility and antigen response in intact lymph node. Science 296(5574):1869–1873
Mitsui H et al (2012) Combined use of laser capture microdissection and cDNA microarray analysis identifies locally expressed disease-related genes in focal regions of psoriasis vulgaris skin lesions. J Invest Dermatol 132(6):1615–1626
Miyake K et al (2017) Trogocytosis of peptide-MHC class II complexes from dendritic cells confers antigen-presenting ability on basophils. Proc Natl Acad Sci U S A 114(5):1111–1116

Mora JR et al (2008) Vitamin effects on the immune system: vitamins A and D take centre stage. Nat Rev Immunol 8(9):685–698
Moyron-Quiroz JE et al (2004) Role of inducible bronchus associated lymphoid tissue (iBALT) in respiratory immunity. Nat Med 10(9):927–934
Mrass P et al (2006) Random migration precedes stable target cell interactions of tumor-infiltrating T cells. J Exp Med 203(12):2749–2761
Mueller SN et al (2014) Tissue-resident T cells: dynamic players in skin immunity. Front Immunol 5:332
Muller AJ et al (2012) CD4+ T cells rely on a cytokine gradient to control intracellular pathogens beyond sites of antigen presentation. Immunity 37(1):147–157
Munoz MA et al (2014) T cell migration in intact lymph nodes in vivo. Curr Opin Cell Biol 30:17–24
Nakajima S et al (2012) Langerhans cells are critical in epicutaneous sensitization with protein antigen via thymic stromal lymphopoietin receptor signaling. J Allergy Clin Immunol 129 (4):1048–1055. e1046
Natsuaki Y et al (2014) Perivascular leukocyte clusters are essential for efficient activation of effector T cells in the skin. Nat Immunol 15(11):1064–1069
Neyt K et al (2012) Tertiary lymphoid organs in infection and autoimmunity. Trends Immunol 33 (6):297–305
Ng LG et al (2011) Visualizing the neutrophil response to sterile tissue injury in mouse dermis reveals a three-phase cascade of events. J Invest Dermatol 131(10):2058–2068
Okada T et al (2016) In vivo multiphoton imaging of immune cell dynamics. Pflugers Arch 468 (11–12):1793–1801
Ono S et al (2018) Requirement of MHC class I on radioresistant cells for granzyme B expression from CD8(+) T cells in murine contact hypersensitivity. J Dermatol Sci
Pitzalis C et al (2014) Ectopic lymphoid-like structures in infection, cancer and autoimmunity. Nat Rev Immunol 14(7):447–462
Randolph GJ (2001) Dendritic cell migration to lymph nodes: cytokines, chemokines, and lipid mediators. Seminars in immunology. Elsevier
Reiss Y et al (2001) CC chemokine receptor (CCR)4 and the CCR10 ligand cutaneous T cell-attracting chemokine (CTACK) in lymphocyte trafficking to inflamed skin. J Exp Med 194 (10):1541–1547
Rubenfeld MR et al (1981) Induction of lymphocyte differentiation by epidermal cultures. J Invest Dermatol 77(2):221–224
Sawada YH, T, Hanakawa S, Nakamizo S, Murata T, Ueharaguchi-Tanada Y, Ono S, Amano W, Nakajima S, Egawa G, Tanizaki H, Otsuka A, Kitoh A, Dainichi T, Ogawa N, Kobayashi Y, Yokomizo T, Arita M, Nakamura M, Miyachi Y, Kabashima K (2015) Resolvin E1 inhibits dendritic cell migration in the skin and attenuates contact hypersensitivity responses. J Exp Med (in press)
Schenkel JM et al (2014) T cell memory. Resident memory CD8 T cells trigger protective innate and adaptive immune responses. Science 346(6205):98–101
Schlitzer A et al (2015) Dendritic cells and monocyte-derived cells: two complementary and integrated functional systems. Seminars in Cell and Developmental Biology. Elsevier
Shin H, Iwasaki A (2013) Tissue-resident memory T cells. Immunol Rev 255(1):165–181
Sigmundsdottir H et al (2007) DCs metabolize sunlight-induced vitamin D3 to 'program' T cell attraction to the epidermal chemokine CCL27. Nat Immunol 8(3):285–293
Soler D et al (2003) CCR4 versus CCR10 in human cutaneous TH lymphocyte trafficking. Blood 101(5):1677–1682
Stingl G et al (1978) Immunologic functions of Ia-bearing epidermal Langerhans cells. J Immunol 121(5):2005–2013
Streilein JW (1978) Lymphocyte traffic, T-cell malignancies and the skin. J Invest Dermatol 71 (3):167–171
Streilein JW (1983) Skin-associated lymphoid tissues (SALT): origins and functions. J Invest Dermatol 80(Suppl):12s–16s

Streilein JW (1985) Circuits and signals of the skin-associated lymphoid tissues (SALT). J Invest Dermatol 85(1 Suppl):10s–13s
Tietz W et al (1998) CD4+ T cells migrate into inflamed skin only if they express ligands for E- and P-selectin. J Immunol 161(2):963–970
Toews GB et al (1980) Langerhans cells: sentinels of skin associated lymphoid tissue. J Invest Dermatol 75(1):78–82
Tomura M et al (2010) Activated regulatory T cells are the major T cell type emigrating from the skin during a cutaneous immune response in mice. J Clin Investig 120(3):883–893
Wakim LM et al (2010) Memory T cells persisting within the brain after local infection show functional adaptations to their tissue of residence. Proc Natl Acad Sci USA 107(42):17872–17879
Wang L et al (2008) Langerin expressing cells promote skin immune responses under defined conditions. J Immunol 180(7):4722–4727
Weninger W et al (2014) Leukocyte migration in the interstitial space of non-lymphoid organs. Nat Rev Immunol 14(4):232–246
Wu T et al (2014) Lung-resident memory CD8 T cells (TRM) are indispensable for optimal cross-protection against pulmonary virus infection. J Leukoc Biol 95(2):215–224
Zaba LC et al (2007) Amelioration of epidermal hyperplasia by TNF inhibition is associated with reduced Th17 responses. J Exp Med 204(13):3183–3194

Memory Lymphocyte Clusters in Genital Immunity: Role of Tissue-Resident Memory T Cells (T_{RM})

Norifumi Iijima

Contents

Abstract Development of front-line defenses in genital tissues is important to inhibit viral/bacterial replication and to eliminate sexually transmitted diseases. In this chapter, we discuss the cellular composition, location, and function of memory lymphocyte clusters deployed in mucosal tissues and compare them with those in secondary lymphoid organs and tertiary lymphoid structures.

Abbreviations

Ab	Antibodies
APC	Antigen-presenting cells
BALT	Bronchus-associated lymphoid tissues

N. Iijima (✉)
National Institutes of Biomedical Innovation, Health and Nutrition, 7-6-8 Saito Asagi, Ibaraki City, Osaka 567-0085, Japan
e-mail: norifumi.iijima@aya.yale.edu

Current Topics in Microbiology and Immunology (2020) 426: 83–118
https://doi.org/10.1007/82_2020_213

Published Online: 29 May 2020

Blimp1	B lymphocyte-induced maturation protein-1
CLA	Cutaneous lymphocyte-associated antigen
DC	Dendritic cells
DLN	Draining lymph nodes
FABP	Fatty-acid-binding proteins
FcRn	Neonatal Fc receptor
FDC	Follicular dendritic cells
FRT	Female reproductive tract
HBV	Human hepatitis B virus
HEV	High endothelial venules
HPV	Human papilloma virus
HSV	Herpes simplex virus
ICAM-1	Intercellular adhesion molecule-1
IFITM3	Interferon-induced transmembrane protein 3
Ig	Immunoglobulin
KLF	Kruppel-like factors
LA	Lactic acid
LC	Langerhans cells
LN	Lymph nodes
LT	Lymphotoxin
LTβR	LT β receptor
LTi	Lymphoid tissue inducer cells
LTo	Lymphoid tissue organizer cells
MAdCAM-1	Mucosal addressin cell adhesion molecule-1
MLC	Memory lymphocyte clusters
NK	Natural killer
pIgR	Polymeric Ig receptor
RANK	Receptor activator of nuclear factor κB
PNAd	Peripheral node addressin
S1PR1	Sphingosine-1-phosphate receptor 1
SLO	Secondary lymphoid organs
STD	Sexually transmitted diseases
T_{CM}	Central memory T cells
TCR	T cell receptor
TD	Thoracic duct
T_E	Effector T cells
T_{EM}	Effector memory T cells
TF	Transcription factor
TGF-β	Transforming growth factor-beta
Th	T helper cells
TK	Thymidine kinase
TLS	Tertiary lymphoid structures
T_M	Memory T cells
TNF	Tumor necrosis factor

Treg	Regulatory T cells
T_{RM}	Tissue-resident memory T cells
VCAM-1	Vascular cell adhesion molecule-1
VEDC	Vaginal epithelial DC
VEGFC	Vascular endothelial growth factor C
WT	Wild-type

1 Introduction

Over the past three decades, various types of lymphoid aggregates have been found in the skin, mucosal tissues, and central nervous system. In contrast to secondary lymphoid organs, lymphoid aggregates in peripheral tissues are not preprogramed in early postnatal life. Instead, chronic inflammation, viral/bacterial infection, the onset of autoimmune diseases, and tumorigenesis can trigger the genesis of lymphoid aggregates containing lymphocytes and stromal cells. These include memory lymphocyte clusters (MLC) and tertiary lymphoid structures (TLS) such as inducible Bronchus-associated lymphoid tissues (iBALT).

In this chapter, we will compare and contrast the cellular composition, distribution, and function of MLC with that of TLS formation and secondary lymphoid organs (SLO) containing lymph nodes (LN). Given that tissue-resident memory T cells (T_{RM}) accumulate in MLC, we will discuss current knowledge regarding the function of T_{RM} retained in peripheral tissues, including in MLCs, and consider the implications of these cells for the development of next-generation medical treatments.

2 Tertiary Lymphoid Structures and Memory Lymphocyte Cluster-like Structures

SLO, including the white pulp of the spleen, LN, the appendix, and mucosal-associated lymphoid tissues are required to initiate the generation of the antigen-specific immunity mediated by T and B cells (van de Pavert and Mebius 2010). In this respect, SLO have all the functionality to maximize the adaptive immune response in terms of the localization of immune cells (Schulz et al. 2016). Stromal cells, antigen-presenting cells (APC) including dendritic cells (DC), follicular DC (FDC), macrophages, and B cells and naïve T cells are strategically deployed in SLO (Fig. 1a) (Randall et al. 2008). In addition, SLO are genetically programmed to be generated in predetermined places within the body (Drayton et al. 2006). During the development of SLO, CXCL13 produced by resident mesenchymal cells initially tempts the precursors of $CD45^{+}$ $CD3^{-}$ $CD4^{+}$ $CXCR5^{+}$

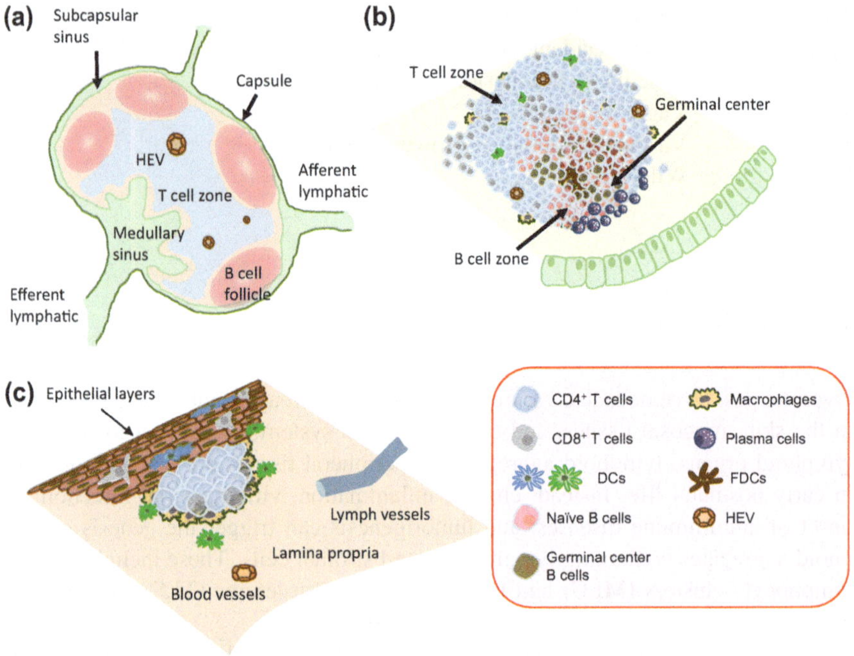

Fig. 1 Schematic representation of LN, TLS, and MLC. **a** LN comprise organized cell populations, including immune cells (T cells in T cell zones and B cells in B cell zones) and stromal cells covered by a fibrous capsule and an underlying subcapsular sinus. In addition, LN contain lymphatic vasculature (afferent and efferent lymphatics) and HEV. **b** The majority of TLS are not encapsulated. Instead, TLS form lymphoid aggregates consisting of T cell zones containing naïve T cells, T_{CM}, DC, macrophages, HEV, and fibroblastic reticular cells and a B cell zone with a germinal center, plasma cells, and FDC. **c** MLCs are formed beneath mucosal epithelial layers with the clusters of $CD4^+$ T_{RM}, $CD8^+$ T_{RM} and MHC class II^+ APC

lymphoid tissue inducer cells (LTi), also known as type 3 innate lymphoid cells bearing RAR-related orphan receptor gamma t to the lymph node anlagen from neighboring blood vessels. Thereafter, the accumulation of receptor activator of nuclear factor κB $(RANK)^+$ LTi facilitates the expression of lymphotoxin (LT) $\alpha 1\beta 2$ and tumor necrosis factor (TNF)-related activation-induced cytokine so that LTi can activate LT β receptor $(LT\beta R)^+$ stromal cells to differentiate into lymphoid tissue organizer cells (LTo) (Chang and Turley 2015). As a consequence of LTβR and RANK activation, LTo attract other hematopoietic cells, mainly lymphocytes through the expression of CCL19 and CCL21 as well as further recruitment of LTi for the development of LN tissue progenitors. Simultaneously, LTo express adhesion molecules including intercellular adhesion molecule-1 (ICAM-1), vascular cell adhesion molecule-1 (VCAM-1), and mucosal addressin cell adhesion molecule-1 (MAdCAM-1) to adhere to infiltrating immune cells (Denton et al. 2019). Thereafter, LTo give rise to various LN stromal subsets, including FDC and marginal reticular cells (Barone et al. 2016). At this time, high endothelial venules (HEV) express MAdCAM-1 but not peripheral node addressin (PNAd). During the maturation of LN, HEV become $PNAd^+$ and down-regulate MAdCAM-1 in the majority of peripheral LN, although in some LN (mesenteric, sacral, cervical), HEV express both PNAd and MAdCAM-1 (Ager 2017; Bistrup et al. 2004; Soderberg et al. 2004).

In some cases, acute inflammation within peripheral tissues results in prolonged production of CXCL13 and IL-7 from stromal cells to attract LTi, which produce IL-17 and LT $\alpha 1\beta 2$ along with the production of TNF-alpha (TNFα) and LTα. In response to the recruitment of LTi to the site of inflammation, the inflammatory response of T helper cell 17 (Th17) cells, B cells, and macrophages is suggested to initiate the generation of TLS in various pathological contexts (Guedj et al. 2014; Lochner et al. 2011; Peters et al. 2011). At the site of inflammation, the TLS initiator cells, including LTi and Th17 cells, interact with $LT\beta R^+$ local stromal cells through $LT\alpha 1\beta 2$ secretion so that stromal cells produce vascular endothelial growth factor C (VEGFC) to promote HEV development in peripheral tissues (Furtado et al. 2007). Likewise, stromal cells also secrete CCL19, CCL21, CXCL12, and CXCL13 and upregulate ICAM-1, VCAM-1, and MAdCAM-1 to recruit lymphocytes and promote TLS structural organization, including a T cell zone with clusters of $TCR\beta^+$ cells and a follicular $CD20^+$ B cell zone (Fig. 1b) (Barone et al. 2016). FDC differentiation from local fibroblasts is especially critical for the development of B cells in TLS (Sautes-Fridman et al. 2019). In these areas, the majority of lymphocytes are $CD62L^+CD44^-$ naïve T cells, naïve B cells, and $CD62L^+CD44^+$ central memory T cells (T_{CM}).

Draining LNs (DLN) are connected with peripheral tissues through afferent lymphatics, providing entry sites for invading pathogens (Fig. 1a). There is no doubt that SLO are indispensable for the induction of robust antigen-specific immunity, while, in general, it takes 5–7 days to generate antigen-specific effector T cells (T_E) that are poised to migrate to the site of infection (Permanyer et al. 2018). Meanwhile, pathogens invading through mucosal tissues are able to replicate and spread rapidly to various organs. In this regard, tertiary lymphoid structures (TLS) in peripheral tissues are capable of immediately challenging invading pathogens. It is well known that TLS-like structures appear in peripheral tissues following local inflammation, viral or bacterial infection, tumor progression or the onset of autoimmune diseases (Corsiero et al. 2019; Fleige et al. 2014; Fridman et al. 2017; Kabashima et al. 2019; Lucchesi et al. 2014; Moyron-Quiroz et al. 2004). Compared with LN, the majority of TLS lack an organized clustering of immune and stromal cells encapsulated by coated layers composed of a fibrous capsule and a subcapsular sinus (Fig. 1b). In most cases, TLS represent a non-encapsulated aggregation of immune and stromal cells confined in an organ or peripheral tissues (Alsughayyir et al. 2017).

In peripheral tissues, MLC-like structures are found in the female reproductive tract (FRT), skin, ocular tissue, and intestine in humans and rodents (Fig. 1c) (Collins et al. 2016; Iijima et al. 2008a; Morrison and Morrison 2000; Reacher et al. 1991; Zhu et al. 2009). In contrast to typical TLS formation, $PNAd^+$ HEV, $CD35^+$ cells, and B cell areas are lacking within these MLC, indicating that naïve T cells, naïve B cells and $CD62^+$ $CD44^+$ T_{CM} are largely absent. Instead, the center of all known MLC consists of $CD11b^+$ and/or CX_3CR1^+ APC (Fig. 1c). In addition, $CD4^+$ T_{RM} mainly form clusters with APC in close proximity to the mucosal epithelium, whereas $CD8^+$ T_{RM} are scattered in the cluster. The majority of $CD103^+$ $CD8^+$ T_{RM} are retained in the epidermis or mucosal epithelium, which is required for signaling via the transforming growth factor-beta (TGF-β) cascade. In contrast, $CD8^+$ T_{RM} in lymphoid clusters located in the intestine express CD103 at low levels; therefore TGF-β signaling pathway is likely to be dispensable for the maintenance of $CD103^-$ $CD8^+$ T_{RM} in the intestinal lamina propria (Bergsbaken and Bevan 2015). Instead of TGF-β-mediated retention control, antigenic stimulation appears to be involved in the residency of $CD103^-$ $CD8^+$ T_{RM}. In the vaginal lamina propria and skin dermis, both IFN-γ from $CD4^+$ T_{RM} and CCL5 from APC are required for the integrity of MLC formation (Collins et al. 2016; Iijima and Iwasaki 2014). However, the mechanisms of MLC development remain unknown, although LTα-mediated pathways are not likely to be involved in the retention of virus-specific T_{RM} in genital tissues following immunization with attenuated HSV (Roth et al. 2013).

3 Protective Immunity in Genital Mucosa

Both the female and male genital tracts are essential for fertilization and pregnancy. Given that the secretion of sex hormones is tightly regulated during the menstrual cycle, immune responses in the genital microenvironment are uniquely affected by hormone control. At the timing of conception, the trophoblast cells of the embryo adhere to the uterine lining and then invade into the maternal uterine decidua (Schatz et al. 2016). Following trophoblast differentiation to form a mature placenta, the immune response at the maternal-fetal interface establishes tolerance over time (Munoz-Suano et al. 2011). Natural killer (NK) cells and APC, including DC and macrophages in the decidua in close proximity to invading trophoblasts and paternally-derived alloantigens, are detected in the developing placental and fetal tissues (Houser et al. 2011). Their interaction triggers a sequence of immune responses that are initiated prior to conception and persist through gestation and delivery of the newborn. Among innate immune cells, $CD56^{hi}$ $CD16^{lo}$ NK cells, which are unique to the uterus, accumulate in the decidua during the progestational phase of the menstrual cycle after implantation is initiated (Koopman et al. 2003). The uterine NK cells are essential for placental development through the provision of growth factors and facilitation of adaptations to the uterine vasculature to help trophoblast invasion (Beaman et al. 2014). In terms of APC, decidual macrophages promote a tolerant microenvironment in healthy individuals through the expression of PD-1 and M2-like characteristics, whereas PD-1 is down-regulated on decidual macrophages that possess the M1 phenotype in patients of recurrent miscarriage (Shimada et al. 2018; Zhang et al. 2018). DC that accumulate in the uterus participate in decidual angiogenesis by secreting soluble fms-like tyrosine kinase-1 and

TGF-β1, which is independent of immunological tolerance (Plaks et al. 2008). Thereafter, Collins MK et al. (2009) demonstrated that DC enriched in the decidua fail to migrate into DLN following exposure to fetal/placental antigens, indicating that the decidual DC play a critical role in blocking immune rejection of the fetus (Collins et al. 2009). In addition to innate immune cells, the adaptive immune response based on regulatory T cells (Treg) and effector/memory T cells is also important for pregnancy tolerance. In particular, to allow reproduction, Treg are necessary to control inflammation in decidual tissues for embryo implantation and progression of gestation (Robertson et al. 2018).

In the human uterine endometrium and cervix, lymphoid clusters containing $CD8^+$ T cells and B cells develop in the stratum basalis of the endometrium during the proliferative phase of the menstrual cycle (Johansson et al. 1999; Wira et al. 2014; Yeaman et al. 2001). Given that the clusters are not found at menses or during gestation, their formation is likely to be under the control of the menstrual cycle (Yeaman et al. 1997). $CD20^+$ B cells are located at the center of lymphoid clusters and are surrounded by $CD45RO^+$ $CD8^+$ T cells and macrophages. Remarkably, these $CD45RO^+$ $CD8^+$ T cells express high levels of CD69 (Yeaman et al. 2001), suggesting that these memory $CD8^+$ T cells are a tissue-resident population. Similarly, structured lymphoid clusters are found in rats and mice and they enlarge to form mural structures by day 10.5 of gestation. As for the development of secondary lymphoid organs, LTα and/or LTβ play essential roles in initiating their organization. LTα is secreted from endometrial stromal cells and transformed decidual cells in human pregnancy (Vince et al. 1992). LTα and LTβ are detected in placental cells, fetal trophoblasts, and decidual macrophages (Phillips et al. 2001). In contrast, the formation of lymphoid clusters in the uterine endometrium does not appear to be affected by the expression of LTα or LTβ, indicating that the clusters are typical tertiary lymphoid structures (Kather et al. 2003; Moyron-Quiroz et al. 2004). Given that the size and maintenance of these lymphoid clusters are influenced by sex hormones, their formation in the uterus indicates involvement from pregnancy to childbirth, although their exact function remains unknown.

For genital tissues to perform their reproductive functions without problems, a local immune response must protect against causative infectious agents of sexually transmitted diseases (STD) (Iwasaki 2010). To this end, genitourinary tracts must deploy highly robust and effective first-line of defenses against viral, bacterial, fungal, and parasitic pathogens, while maintaining tolerance to allo-immune responses against spermatozoa and the semi-allogeneic fetus (Wira et al. 2014).

In contrast to other mucosal tissues, the immune defense system in the FRT is affected by its distinctive features of anatomy, hormonal control, mucus, microbiota, lymphatics, and cellular composition (Deruaz and Luster 2015; Iwasaki 2016; Wira et al. 2015). In terms of vaginal microbiota, the FRT is dominated by *Lactovacillus* species, which may play a role in protection against HIV-1 transmission (Chen et al. 2017). In addition, reduced colonization of vaginal lactobacilli is closely associated with the onset of opportunistic diseases and an increased risk of HIV-1 infection (Quinn and Overbaugh 2005). As a potential mechanism of protection mediated by *Lactovacillus* species, vaginal *Lactobacillus* spp. is known to secrete antimicrobial factors, such as bacteriocins and organic acid metabolites, mainly lactic acid (LA). Protonated LA levels in cervicovaginal fluid inactivate HIV-1 replication; while the exact mechanism of this remains unclear, and it is not likely to be mediated by lactate anions or pH (Tyssen et al. 2018). Microbiota, including vaginal lactobacilli, has distinct types of immunoregulatory properties that have an impact on unique characteristics and functions of resident immune cells (Song et al. 2018).

To drive the robust protective immunity mediated by antigen-specific T and B cells, tissue-resident DC have highly potent antigen-presenting capacity to initiate the antigen-specific immune response (Iwasaki 2007). Therefore, the phenotypic and functional signatures of DC in genital tissues determine the quality of immune protection. The phenotypic features of DC in genital tissues are quite distinct from those of DC in other tissues (Iijima et al. 2008b; Parr et al. 1991). In particular, vaginal epithelial DC (VEDC) consist of several distinct populations in mice (Iijima et al. 2007) and humans (Ballweber et al. 2011; Duluc et al. 2013), whereas Langerhans cells (LC) ($CD207^{hi}$ $CD326^{+}$ $CD11c^{+}$ $MHCII^{+}$ DC) are predominantly located in skin epidermis (Merad et al. 2002). Likewise, dendritic cell populations in ocular and oral epithelia can be segregated into at least two distinct populations based on the expression of CD11b and CD103 ($F4/80^{-}$) (Capucha et al. 2015; Hattori et al. 2011). These findings suggested the fate of dendritic cell differentiation depending on whether stratified squamous epithelia is cornified or not. In fact, in adult skin epidermis, LC are not replenished by circulating BM-derived precursors (Hovav 2018). Instead, Langerhans cell precursors enter the epidermis during embryonic development and then self-renew there (Chorro et al. 2009; Hoeffel et al. 2012). During early embryogenesis, yolk sac macrophages give rise to LC. In addition, LC are likely to arise from liver monocytes (Hoeffel et al. 2012). In contrast, VEDC develop from circulating BM-derived precursors, including monocytes, common dendritic cell progenitors, and pre-DC (Capucha et al. 2015; Iijima et al. 2007). As another hallmark of DC in epithelia, skin, oral, and ocular DC expresses high levels of CD207, whereas VEDC express CD207 at low levels (Capucha et al. 2015; Duluc et al. 2013), suggesting that CD207 expression in vaginal tissues is influenced by unknown genital-specific factors. These findings indicate that the final differentiation of genital DC is determined by the properties of genital tissues. As for the function of DC in genital tissues, submucosal DC in the vaginal lamina propria but not VEDC are required for the generation of Th1 cells in DLN (Zhao et al. 2003), whereas VEDC appear to be involved in the differentiation of Th17 cells in DLN (Anipindi et al. 2016; Hervouet et al. 2010). To control the migration of these DC, CTLA4 expressed on $FoxP3^{+}$ Treg is involved in the recruitment of vaginal DC into DLN to initiate antigen-specific T cell responses (Lund et al. 2008; Soerens et al. 2016). In addition, $CD8\alpha^{+}$ and $CD8\alpha^{-}$ DC in DLN present viral antigens to $CD8^{+}$ T cells to differentiate into IFN-γ producing cells, whereas the origin of DC in vaginal tissues remains elusive (Lee et al. 2009). After migratory DC leave genital tissues, rapid recruitment of $Ly6C^{hi}$ monocytes from the blood occurs in vaginal tissues (Iijima et al. 2007). Thereafter, $Ly6C^{hi}$ monocytes capture viral antigens and upregulate MHC class II and costimulatory molecules (Iijima et al. 2011). Furthermore, the monocyte-derived APC are able to present antigens to effector Th1 cells that subsequently enter into vaginal tissues. In addition, the reactivated Th1 cells are critical for inhibiting viral replication in the vaginal epithelium (Iijima et al. 2011). With respect to the function of other dendritic cell subsets, topical application of aminoglycoside antibiotics (neomycin) to vaginal tissues triggers the recruitment of $CD103^+$ $XCR1^+$ DC, which is responsible for a significant upregulation of interferon-stimulated genes in vaginal tissues. This contributes to the protection against vaginal HSV-2 and Zika virus infection in a microbiota-independent manner (Gopinath et al. 2018).

More than 30 pathogens cause STD. With regard to bacterial infection, *Neisseria gonorrhoeae*, *Chlamydia trachomatis*, *Treponema pallidum,* and *Trichomonas vaginalis* lead to reproductive failure and possible death if untreated. In terms of virus infection, human papillomavirus (HPV) infection causes cervical cancer. In addition, HIV and genital herpes cause chronic infection and are difficult to completely cure. Thus, safe and effective vaccines to prevent sexually transmitted infections are urgently needed. However, vaccines are currently only available for two of pathogens causing STD (HPV and hepatitis B virus (HBV)), despite many of vaccine trials having been conducted. Protection against HPV infection by the prophylactic vaccine is thought to be largely mediated by antibodies (Ab); however, the control of pathogens that replicate intracellularly is likely to require cellular immunity, requiring the development of therapeutic vaccines. To block invading pathogens binding with host cells, including epithelial cells, Ab must be secreted from the mucosal lamina propria into the lumen. In FRT, the upper genital tract is classified as a type I mucosal surface, which consists of simple columnar epithelium. On type I mucosal epithelial cells, both neonatal Fc receptor (FcRn), which plays a prominent role in the transport of immunoglobulin G (IgG) into the lumen, and polymeric Ig receptor (pIgR), which binds dimeric IgA to transport it into the lumen, are expressed. However, the lower genital tissues, including the ectocervix and vagina, are classified as type II mucosal surfaces and are covered by stratified squamous epithelial cells. On type II mucosal epithelia, FcRn but not pIgR is expressed, indicating that IgG but not IgA is the main effector molecule inhibiting the attachment of infectious agents to epithelial cells in the lower reproductive tract. For IgG to exert a protective function against invading pathogens, Igs in the FRT must be adequately delivered to the mucosal lumen. In addition, Ig levels in the lumen of genital tissues are influenced by sex hormones. (Wira et al. 2015). In fact, IgG levels in secretions from uterine tissues are much higher than those in the fallopian tube during the periovulatory phase (Safaeian et al. 2009a). Furthermore, following HPV vaccine immunization, transition to the mid-cycle during ovulatory cycles causes a dramatic decrease in IgG levels that react to HPV virus-like particles in cervical secretions (Nardelli-Haefliger et al. 2003). Collectively, unique immune defense systems that adapt to the estrous cycle are strategically deployed in genital tissues.

It is currently unclear whether hormonal control affects the efficacy of HPV vaccines, but licensed prophylactic vaccines against HPV infection are highly effective in preventing the infection (Joura et al. 2015). In contrast, no vaccine trials for genital herpes have succeeded in blocking HSV-2 infection in vaginal tissues. Following HPV vaccines or HSV vaccine candidate immunization, high titers of antigen-specific Igs are maintained in peripheral blood for over years and long-term antibody responses have also been observed (Belshe et al. 2012; Slifka and Amanna 2019). Contrary to serum antibody responses, IgA and IgG levels in cervical tissues are not associated with levels in blood (Safaeian et al. 2009b). In addition, Ig levels in genital tissues are markedly lower than those in peripheral blood, indicating that passive transfer of IgA and IgG thorough pIgR and FcRn into the lumen of genital tissues does not completely explain the mechanism of Ig transport. To establish HPV infection, the virus must reach epithelial basal cells with stem cell-like properties at the bottom of an epithelium. Following wound or epithelial trauma, HPV viral particles are able to access to the epithelial stem cells (Egawa et al. 2015). HSV infection, however, does not need epithelial trauma to access epithelial basal cells because it directly infects vaginal epithelial cells through viral entry receptors, including nectin-1 (Linehan et al. 2004). Recently, Oh JE et al. (2019) clearly demonstrated that the levels of antigen-specific Ab are significantly increased in the murine vaginal lumen following epithelial barrier breach (Oh et al. 2019), indicating that following epithelial trauma, leakage of HPV-specific Ab into the lumen blocks the establishment of HPV infection in individuals who received a prophylactic HPV vaccine. Furthermore, the mouse model evidence explains why the robust antibody responses induced by a prophylactic for HSV-2 vaccine are entirely ineffective at preventing the infection.

4 Generation and Function of T_{RM} in Peripheral Tissues

Following immunization or exposure of vaginal tissues to a pathogen, DC localized in the vaginal epithelium and lamina propria capture the antigen and migrate into DLN to generate antigen-specific T_E (Deruaz and Luster 2015; Iwasaki 2010). Once T_E migrate from DLN to the infection site, including the mucosa, they have the intrinsic capability to combat a replicating pathogen by means of cytotoxic activity and cytokine production. However, after the clearance of an invading pathogen or a few weeks after immunization, T_E mostly lead to apoptosis during a contraction phase (Badovinac et al. 2002). This process is likely to be preprogrammed to cease an undesirable inflammatory response and tissue damage mediated by T_E. However, a small number of T_E differentiate into long-lived memory T cells (T_M). T_M are able to rapidly exert a powerful effector function based on cytokine production and cytotoxic activity following reinfection with the same pathogen. This hallmark of T_M substantially contributes to combat against a secondary encounter with the same pathogen. Therefore, the generation of T_M following primary infection with a pathogen or immunization with a vaccine is an important component in achieving long-term protection and developing effective therapeutics. Two distinct types of T_M are generated: central memory T cells (T_{CM}) and effector memory T cells (T_{EM})

(Sallusto et al. 1999). T_{CM} express LN homing receptors, including CCR7 and CD62L, so that T_M can access to LN, which consist of a network of $CCL19^+$ stromal cells and $PNAd^+$ endothelial cells. Given their phenotypic features, T_{CM} are able to circulate between the blood and LN by entering the efferent lymph and returning to the circulation via the thoracic duct (TD) (Sigmundsdottir and Butcher 2008). In contrast, T_{EM} barely express CCR7 and CD62L. Instead, T_{EM} express tissue-homing receptors including $\alpha4\beta7$ and CCR9 (for intestine), $\alpha1\beta1$ (VLA-1), CXCR3 and CXCR6 (for lung tissues), CXCR3 and CCR5 (for genital tissues), $\alpha4\beta1$ (VLA-4), CXCR3 and CCR6 (for neuronal tissues) and E-selectin ligands, P-selectin ligands, CCR4, cutaneous lymphocyte-associated antigen (CLA) and CCR8 (for skin tissues) (Bromley et al. 2008; McCully et al. 2012). Accordingly, it is conceivable that almost all T_{EM} continuously circulate between the blood and peripheral tissues. To elucidate the mechanism by which T_{EM} move from tissues, Bromley SK et al. (2005) and Debes GF et al. (2005) demonstrated that CCR7 expressed on T cells is involved in their exit from peripheral tissues, indicating that CCR7 might be upregulated on T_{EM} (Bromley et al. 2005; Debes et al. 2005). However, in various experimental settings, including pathogen infection, asthma and autoimmune diseases, $CD44^{hi}$ $CD62L^-$ T_E in peripheral tissues are $CCR7^{low}$ and direct evidence of continuous circulation was nearly lacking (Jameson and Masopust 2018; Masopust and Soerens 2019). Similarly, sphingosine-1-phosphate receptor 1 (S1PR1) guides naïve $CD4^+$ T cells but not activated T cells toward afferent lymphatic vessels (Ledgerwood et al. 2008). These data indicate that S1PR1 rather than CCR7 signaling might regulate the movement of T cells to the afferent lymphatics from tissues. However, the detailed mechanism by which T cells enter the lymphatic system remains to be fully elucidated. As such, recent findings of $CD8^+$ T_M subsets based on the expression of CX_3CR1 provide a more feasible classification of T_M to extend the traditional concept regarding T_{CM} and T_{EM}. Gerlach C et al. identified three distinct populations of murine T_M: CX_3CR1^{lo} $CXCR3^+$ $CD8^+$ T_{CM}, CX_3CR1^{int} $CXCR3^+$ $CD8^+$ T_M, and CX_3CR1^{hi} $CXCR3^-$ T_{EM} (Gerlach et al. 2016). Both CX_3CR1^{lo} $CXCR3^+$ $CD8^+$ T_{CM} and CX_3CR1^{int} $CXCR3^+$ $CD8^+$ T_M are $CD62L^+$ $CD27^+$; therefore, that they can access LN. In contrast, surprisingly, CX_3CR1^{hi} $CXCR3^-$ T_{EM} are not found in peripheral tissues. Instead, CX_3CR1^{int} $CXCR3^+$ $CD8^+$ T_M are able to survey peripheral tissues. Furthermore, CX_3CR1^{int} $CXCR3^+$ $CD8^+$ T_M are responsive to CCL19 and are detected in TD lymph regardless of CD62L expression, indicating that this population is a genuine continuously circulating T_M (Gerlach et al. 2016). In human blood, CX_3CR1^+ $CD62L^-$ and $CX_3CR1^+CD62L^+$ $CD8^+$ T_M are also characterized (Bottcher et al. 2015). Both have the capability to secrete granzyme B and perforin, which are identical to T_{EM}, while CX_3CR1^- $CD62L^+$ $CD8^+$ T_M are able to produce a high level of IL-2 which are analogous to T_{CM}. As for continuously circulating T_M, $CD69^-$ $CD8^+$ T_M, which resemble CX_3CR1^{int} $CXCR3^+$ $CD8^+$ T_M, recirculate between skin tissues and blood under the control of TGF-β activation through keratinocytes expressing $\alpha_v\beta6$ and $\alpha_v\beta8$ (Hirai et al. 2019). In peripheral tissues, approximately 10%-20% of T_M are CX_3CR1^{int} $CXCR3^+$ $CD8^+$ T_M, and the remaining population represents CX_3CR1^- $CD69^+$ $CD62L^-$ $CD103^{+/-}$ $CD8^+$ T_M, which are distinct from T_{EM} and T_{CM} (Gerlach et al. 2016).

Over the past decade, many groups have tried to clarify whether T_M detected in the peripheral tissues are continuously circulating cells or tissue-resident populations (Clark et al. 2012; Masopust et al. 2001). To answer this question experimentally, a parabiosis technique involving the sharing of blood circulation between two mice confirmed that a distinct subset of T_M representing the $CD44^+$ $CD62L^-$ $CD69^+$ $CD103^{+/-}$ population are T_{RM} that are largely disconnected from blood circulation (Iijima and Iwasaki 2014; Jiang et al. 2012; Klonowski et al. 2004; Schenkel et al. 2013; Teijaro et al. 2011). Furthermore, using photoconversion with the parabiosis technique, Park et al. (2018a) elegantly demonstrated that a minority of non-photoconverted circulating $CD4^+$ T_M, $CCR7^+$ $CD62L^-$ $CD69^-$ cells is completely segregated from $CD4^+$ T_{RM} in the skin dermis following *Candida albicans* infection (Park et al. 2018a), which is consistent with a similar population of previously reported circulating CX_3CR1^{int} $CXCR3^+$ T_M (Bromley et al. 2005; Debes et al. 2005; Gerlach et al. 2016; Hirai et al. 2019).

In human tissues, based on phenotypical and functional features of T_{RM}, T_M with a similar phenotype to mouse T_{RM} have been observed (Clark 2015). In a study of patients with genital herpes, HSV-specific CD8αα memory T cells were identified at the dermal-epidermal junction for a long time following resolution of HSV reactivation (Zhu et al. 2013). Furthermore, following alemtuzumab treatment (an anti-CD52 antibody), Watanabe R et al. (2015) elegantly demonstrated that a large number of T cells with an effector function analogous to that of T_{RM} remained in the skin of leukemic cutaneous T cell lymphoma patients despite complete depletion of all circulating T cells (Watanabe et al. 2015). These data directly confirmed the existence of long-lived T_{RM} in human peripheral tissues. In contrast to long-term resident T_{RM}, a minority population of $CD103^+$ CLA^+ $CD69^+$ $CD4^+$ T_{RM} down-regulates CD69 and then transits from the skin to the blood circulation (Klicznik et al. 2019). In particular, $CD103^+$ CLA^+ $CD4^+$ T_{RM} have the capability to produce IL-22 and IL-13 but not IL-17, IFN-γ, IL-4 and GM-CSF, indicating that control of retention is largely mediated by cytokine production from $CD4^+$ T_{RM}.

T_{RM} are mainly found to be seeded in non-lymphoid tissues, particularly at barrier tissues such as mucosal tissues and skin. As tissue-resident T cell populations, $CD8^+$ T_{RM} and $CD4^+$ T_{RM} are found in a variety of peripheral tissues, while their location, differentiation process, phenotypical, and functional features differ as described below.

4.1 The Role of $CD8^+$ T_{RM} in Barrier Tissues

Naïve $CD8^+$ T cells have the capacity to migrate the blood circulation, lymphoid tissues (LNs, spleen and Peyer's patches), and return to the blood via the efferent lymph. In contrast, $CD8^+$ T_M ($CD44^{hi}$, $CD62L^{lo}$) are distributed in peripheral tissues, especially, in the epithelia but not in lamina propria or dermis, long after the clearance of a pathogen (Masopust et al. 2001). Based on their phenotypic characteristics, the majority of $CD8^+$ T_{RM} in skin epithelium, vaginal epithelium, lung tissues (airway, lung interstitium, and upper respiratory tract), brain, salivary grand, liver, and kidney express high levels of CD103 and CD69 (Masopust and Soerens 2019). In particular, CD103 is required for $CD8^+$ T_{RM} retention in skin epidermis (Mackay et al. 2013), and is likely to be involved in the survival of $CD8^+$ T_{RM} in brain tissue following virus infection (Wakim et al. 2010). CD103 appears to interact with E-cadherin expressed on epithelial cells; therefore, CD103 would be essential for the retention of $CD8^+$ T_{RM} in squamous epithelia (type II mucosa) and monolayer epithelium (type I mucosa) but not in brain tissues. To differentiate into $CD8^+$ T_{RM}, T_{RM} precursors in blood must enter peripheral tissues following some sort of stimulus. Topical application of CXCR3 ligands or nonspecific inflammatory stimuli were able to tempt T_{RM} precursors from the blood into peripheral tissues (Mackay et al. 2012; Shin and Iwasaki 2012); therefore, both $CXCR3^{hi}$ $CX3CR1^-$ T_{CM} and $CXCR3^{hi}$ $CX3CR1^{int}$ T_M have the potential to differentiate into T_{RM}. Following transit into peripheral tissues, TGF-β signaling can control the expression of CD103, which is indispensable for the maintenance of $CD8^+$ T_{RM} in skin epidermis and gut tissues (Casey et al. 2012; Mackay et al. 2013; Sheridan et al. 2014; Skon et al. 2013; Zhang and Bevan 2013). Furthermore, Mohammed J et al. (2016) demonstrated that retention of $CD8^+$ T_{RM} requires $\alpha_v\beta6$ and $\alpha_v\beta8$ expressed on keratinocytes through latent TGF-β activation in the epidermis (Mohammed et al. 2016). In the intestine, $\alpha_v\beta6$ controls the residence of $CD8^+$ T_{RM}, indicating that $\alpha_v\beta6$ plus $\alpha_v\beta8$ or $\alpha_v\beta6$ are necessary for $CD8^+$ T_{RM} to reside in type II mucosa or type I mucosa, respectively.

With regard to factors for $CD8^+$ T_{RM} survival, the expression of Bcl2 is significantly increased in $CD8^+$ T_{RM} in skin epidermis (Park et al. 2018b) and intestine (Bergsbaken and Bevan 2015) indicating long-term persistence in tissue. Of note, Wakim LM et al. (2013) demonstrated that $CD8^+$ T_{RM} maintain a high level of interferon-induced transmembrane protein 3 (IFITM3) expression to survive in lung parenchyma following influenza infection (Wakim et al. 2013). In skin epidermis, IL-15 produced by radio-resistant cells is also essential for the long-term survival of $CD8^+$ T_{RM} (Adachi et al. 2015; Mackay et al. 2015). Furthermore, fatty-acid-binding proteins 4 and 5 (FABP4 and FABP5) and P2RX7 play pivotal roles in the maintenance and function of long-lived $CD8^+$ T_{RM} mediated by regulating oxidative metabolism through exogenous free fatty acid intake (Pan et al. 2017) or by promoting mitochondrial homeostasis (Borges da Silva et al. 2018). However, it is unclear whether Bcl2, IFITM3, IL-15, FABP4, FABP5, and P2RX7 share commonalities to control the survival of $CD8^+$ T_{RM} in mucosal tissues and skin.

In addition to $CD103^+$ $CD8^+$ T_{RM}, long-lived $CD103^-$ $CD8^+$ T_{RM} have been found in intestine (Bergsbaken and Bevan 2015), brain (Steinbach et al. 2016), liver (Mackay et al. 2016), and secondary lymphoid organs (Beura et al. 2018; Schenkel et al. 2014b). Although $CD103^+$ $CD8^+$ T_{RM} need keratinocyte-mediated TGF-β activation for their retention, $CD103^-$ $CD8^+$ T_{RM} appear to require interaction with $CD4^+$ T cells and CX_3CR1^+ APC, which form clusters in the lamina propria of the intestine (Bergsbaken and Bevan 2015), suggesting that the recognition of cognate antigen is involved in the retention. This is in contrast to $CD103^+$ $CD8^+$ T_{RM}, which maintain the residency in a local antigen-independent manner.

As is the case for CD103, upregulation of CD69 in $CD8^+$ T_{RM} has also been reported in various tissues. At the beginning of an immune response, CD69 is an early activation marker that is triggered by type-I interferon signaling in DLN. Upon the upregulation of CD69 on activated T cells, receptor sphingosine 1-phosphate receptor 1 (S1PR1) is down-regulated so that these T cells are retained in DLN and differentiate into T_E (Matloubian et al. 2004). Once T_E move to the blood circulation, the level of CD69 expression is low, whereas, following tissue-resident memory T cell differentiation in tissues, CD69 is re-expressed on T_{RM}. Skon et al. (2013) elegantly elucidated that repression of S1PR1 and the transcriptional factor, KLF2 (Kruppel-like factors 2), which controls expression of S1PR1, is required for the establishment of $CD8^+$ T_{RM} (Skon et al. 2013), indicating that downregulation of S1PR1 expression is also critical for $CD8^+$ T_{RM} generation. However, in various types of experimental setting, including influenza infection, chronic virus infection and protein immunization, the requirement of CD69 for the maintenance of $CD8^+$ T_{RM} varies according to the type of peripheral tissue (Beura et al. 2018). This indicates that other factors also control the expression of KLF2 and S1PR1 for retention (Walsh et al. 2019).

KLF2 downregulation has been extensively recognized as a hallmark of T_{RM}. Recently, homolog of B lymphocyte-induced maturation protein-1 (Blimp1) in T cells (Hobit) was identified to be remarkably upregulated in skin-resident $CD8^+$ T_{RM}, NKT cells and liver-resident NK cells (Mackay et al. 2016). Both Hobit and Blimp-1 control the expression of KLF2, S1PR1, CCR7, CD69, and cytotoxic molecules including granzyme B and TRAIL. In contrast to skin epidermis, Blimp-1 rather than Hobit is required for the establishment of $CD8^+$ T_{RM} in lung tissues (Behr et al. 2019). Furthermore, in human, the level of Hobit expression in T_M in blood is higher than that in T_{RM} in brain (Smolders et al. 2018). With regard to other transcription factors related to the control of $CD8^+$ T_{RM}, RUNX3 also regulates multiple targets that influence the retention of $CD8^+$ T_{RM} (Milner et al. 2017). Hence, $CD8^+$ T_{RM} development might be regulated by an intricate network of transcription factor expression that is influenced by each tissue microenvironment.

An intriguing feature of T_{RM} is their ability to exert a rapid effector function following re-encounter with the same pathogen at the site of infection. The presence of $CD8^+$ T_{RM} facilitates the clearance of invading pathogens, including viruses, bacteria, and fungi. Rapid secretion of cytokines (IFN-γ and TNF-α) and cytotoxic molecules, including granzyme B and perforin, directly blocks the replication of previously encountered pathogen in infected cells. T_{RM} therefore contribute substantially to prevent the spread of infectious agents (Mackay et al. 2013). Furthermore, immediate production of IFN-γ and TNF-α reinforces innate immunity, including DC maturation and NK cell activation, and triggers robust production of chemokines and rapid recruitment of leukocytes from blood circulation (Schenkel et al. 2014a; Schenkel et al. 2013). Collectively, $CD8^+$ T_{RM} are able to orchestrate immediate robust protective immunity as a front-line of defense.

The topical application of chemokines including CXCL9, surfactants including nonoxynol-9 or local antigen expression, recruits circulating $CD8^+$ T cells to mucosal tissues, leading to their retention as $CD8^+$ T_{RM} in the epithelium and lamina propria (Cuburu et al. 2019; Mackay et al. 2012; Shin and Iwasaki 2012; Takamura et al. 2016). These findings have important implications for the future development of effective mucosal vaccines.

4.2 *The Role of $CD4^+$ T_{RM} in Peripheral Tissues*

Over the last few decades, the existence of $CD4^+$ T cells in addition to $CD8^+$ T cells in local tissues has been observed (Reinhardt et al. 2001). At the same time, tissue-tropic T_{EM}-like cells circulate in blood until they are recruited to site of inflammation, while non-inflamed local tissues contain very few T cells. T cells that had migrated into tissues during infections were thought to either exit the tissue or undergo apoptosis after clearance of the infection. However, $CD4^+$ T cells and $CD8^+$ T cells that accumulate at the site of virus infection or protein immunization have been observed (Masopust et al. 2001; Reinhardt et al. 2001). In addition, activated $CD4^+$ T cells ($CD25^+$ and $CD69^+$) persist in lung tissues for several months following virus infection (Hogan et al. 2001). Furthermore, $CD4^+$ T cells that enter into peripheral tissues barely proliferated (Reinhardt et al. 2003), indicating that $CD4^+$ T cells that settle within peripheral tissues have a unique system to control their retention (Schenkel and Masopust 2014).

Similar to $CD8^+$ T_{RM}, $CD4^+$ T_{RM} localize to non-lymphoid tissues, such as skin and mucosal tissues of the lung, small intestine, and FRT for a long period of time (Collins et al. 2016; Glennie et al. 2015; Iijima and Iwasaki 2014; Iijima et al. 2008a; Stary et al. 2015; Steinfelder et al. 2017; Teijaro et al. 2011; Turner et al. 2018; Wilk et al. 2017). As is the case of $CD8^+$ T_{RM}, the majority of $CD4^+$ T_{RM} express high levels of CD69, but CD103 expression varies according to tissue. Contrary to $CD8^+$ T_{RM} that are retained in skin epidermis or mucosal epithelia, the majority of $CD4^+$ T_{RM} are distributed in the skin dermis or mucosal lamina propria, suggesting that the retention mechanism of $CD4^+$ T_{RM} is distinct from that of $CD8^+$ T_{RM}. TGF-β signaling contributes to the maintenance of $CD8^+$ T_{RM} in the epithelial layer, whereas, for IFN-γ^+ $CD4^+$ T_{RM}, the formation of clusters with other resident immune cells, including $CD8^+$ T cells and APC such as macrophages and DC, is required for the retention in the lamina propria of the FRT, lung, and skin dermis (Acosta-Ramirez et al. 2016; Collins et al. 2016; Iijima and Iwasaki 2014). In contrast, IL-17^+ $CD4^+$ T_{RM} do not form clusters in skin dermis following *Candida albicans* infection but these cells are retained in the superficial dermis (Park et al. 2018a). Following skin HSV infection, the localization of IFN-γ^+ $CD4^+$ T_{RM} is, however, confined to lymphocyte clusters formed around hair follicles (Collins et al. 2016). Therefore, each $CD4^+$ T_{RM} subset is suggested to have its own retention-control system in peripheral tissues.

Substantial advances have been made in understanding the differentiation of $CD8^+$ T_{RM}; however, the detailed molecular mechanism of $CD4^+$ T_{RM} development remains relatively elusive. Given the existence of functionally distinct effector Th subsets, such as Th1, Th17, and Th2, identification of $CD4^+$ T_{RM} precursors is much more complicated. Furthermore, both $CD4^+$ T_E and $CD4^+$ T_M exhibit substantial plasticity and easily convert to different linages (Hegazy et al. 2010). Similarly, in $CD8^+$ memory T cell development, $KLRG1^{hi}$ $CD127^{lo}$ cells were identified as a terminal effector population, while the $KLRG1^{lo}$ $CD127^{hi}$ subset contains memory precursor cells with the potential to differentiate into long-lived $CD8^+$ T_M (Joshi et al. 2007). In contrast, Marshall HD et al. (2011) clearly demonstrated that the expression of CD127, the IL-7 receptor α chain, does not characterize memory precursors from $CD4^+$ T_E (Marshall et al. 2011). Instead, $PSGL1^{hi}$ $Ly6C^{lo}$ $T\text{-}bet^{int}$ $CD4^+$ T_E become long-lived T_M, such as $CCR7^{hi}$ $CD62L^{hi}$ T_{CM}. Interestingly, the gene expression level of CX_3CR1 in the $PSGL1^{hi}$ $Ly6C^{lo}$ $T\text{-}bet^{int}$ population is significantly lower than that in the $PSGL1^{hi}$ $Ly6C^{hi}$ $T\text{-}bet^{hi}$ population, which resembles T_{EM}-like cells, indicating that the definition of $CD4^+$ T_M subsets can also be classified by CX_3CR1 expression. It still remains unclear whether the precursor for $CD4^+$ T_{RM} originates from T_{CM}, T_{EM}-like cells or another type of memory T cell. A recent finding clearly demonstrated that effector Th17 cells reached in lung tissues differentiate into $CD4^+$ T_{RM} that have the capability to produce IFN-γ but not IL-17 (Amezcua Vesely et al. 2019), although it remains unknown whether all $CD4^+$ T_{RM} originate from effector Th17 cells.

For long-term survival of $CD4^+$ T_{RM} in peripheral tissues, $CD4^+$ T_{RM} must receive survival signals within the local microenvironment during their generation and differentiation. The common gamma-chain cytokines, including IL-2, IL-7 and IL-15 play pivotal roles in the generation of $CD4^+$ T_M. Among them, IL-2 receptor signaling is necessary for the generation of $CD4^+$ T_{RM} in lung tissues following the induction of allergic asthma or lymphocytic choriomeningitis virus infection (Hondowicz et al. 2016, 2018). The generation of $CD4^+$ T_{RM} through an IL-2-independent pathway has also been reported following lung influenza infection. Besides, IL-2-independent $CD4^+$ T_{RM} require IL-15 for T cell activation, whereas IL-15 was redundant for the maintenance of $CD4^+$ T_{RM} (Strutt et al. 2018). The maintenance of long-lived $CD4^+$ T_M, however, requires IL-7 signaling because IL-7R is highly expressed on naïve $CD4^+$ T cells and $CD4^+$ T_M (Seddon et al. 2003). In a skin model of contact hypersensitivity, $CD4^+$ T_M in skin tissues fail to be retained following ablation of IL-7 from epidermal keratinocytes, indicating that IL-7 signaling in the local microenvironment is essential for the entry or survival of $CD4^+$ T_M (Adachi et al. 2015). Likewise, following lung *Klebsiella pneumonia* infection, $CD4^+$ T_{RM} derived from Th17 cells express high levels of CD127 so that the cells can be maintained in lung tissue through IL-7-mediated signaling (Amezcua Vesely et al. 2019). In contrast, the expression of CD127 on $CD4^+$ T_{RM} is comparable to that on circulating $CD4^+$ T_{EM} in Peyer's patches (Ugur et al. 2014). Similarly, the expression of CD127 on $CD4^+$ T_{RM} was significantly lower than that on circulating $CD4^+$ T_{EM} in the FRT (Beura et al. 2018). Furthermore, a parabiosis study demonstrated that the level of CD122 expression on $CD4^+$ T_{RM} is almost identical to that on circulating $CD4^+$ T_{EM} in the FRT (Beura et al. 2018), suggesting that IL-15 signaling is also dispensable for the retention of $CD4^+$ T_{RM} in genital tissues. Thus, the requirement of IL-7 and IL-15 for the survival of $CD4^+$ T_{RM} appears to be dependent on the properties of peripheral tissues. Collectively, the inflammatory status and immune responses in tissue microenvironments are more likely to shape the retention and function of $CD4^+$ T_{RM} compared with $CD8^+$ T_{RM}.

For this reason, the topical application of any product is not able to generate MLC formation containing $CD4^+$ T_{RM}. Currently, charge-switching synthetic adjuvant particles containing UV-inactivated *Chlamydia trachomatis* is the only topically applied artificial compound that has generated $CD4^+$ T_{RM} in the uterine mucosa (Stary et al. 2015). This finding has profound implications for the development of other vaccines, although it remains unknown whether the retention of $CD4^+$ T_{RM} themselves is sufficient to inhibit the spreading of invading pathogens.

5 Generation and Maintenance of MLC in Genital Tissues

For almost two decades, lymphocyte aggregates, memory lymphocyte clusters (MLC), have been reported in genital tissues following HSV-2 or *Chlamydia trachomatis* infection; however, the exact function of MLC remained elusive until fairly recently (Gillgrass et al. 2005; Morrison and Morrison 2000). In contrast to TLS, $CD62L^{hi}$ $CD44^{lo}$ naïve T cells are not found within the clusters of $CD4^+$ T cells, $CD8^+$ T cells and APC. Similarly, the expression of PNAd, CD31, lymphatic vessel endothelial hyaluronan receptor 1 and CD21/CD35 as the basal components of TLS is not detected within MLC (Fig. 1c) (Table 1), indicating that MLC are not the sites for induction of antigen-specific effector T cells from naïve T cells (Iijima and Iwasaki 2014, 2015). Furthermore, the expression of CCL19, CCL21, CXCL12, and CXCL13 is not found in MLC. Instead, CXCL9 and CCL5 were detected after MLC formed in vaginal tissues (Iijima and Iwasaki 2014) (Table 1). No MLC-like structure has been demonstrated to be affected by sex hormones, except for lymphoid aggregates centered around $CD20^+$ B cells in the uterine endometrium.

At present, MLC are mainly observed in the lamina propria beneath vaginal epithelial layers, which is the entry site of the HSV-2 virus. However, *Chlamydia*-induced MLC are found in vaginal tissues, cervix, and fallopian tubes. Thus, the mechanism of MLC generation and maintenance following HSV-2 infection and *Chlamydia trachomatis* infection is shown separately as follows.

Table 1 Components, immune cells, and fibroblastic cells in secondary lymphoid organs, tertiary lymphoid structures, or memory lymphocyte clusters

	Encapsulated	T cell zone	B cell zone	HEV	Naive T and B cells	T_{CM}	T_{RM}	FDC	Chemokines and cytokines
SLO	+	+	+	+	+	+	+	+	LT, IL-17, CCL19, CCL21, CXCL13
TLS	−	+	+	+	+	+	N/D	+	IL-17, CCL19, CCL21, CXCL13
MLC	−	+	−	−	−	−	+	−	CCL5, CXCL9

N/D not determined
SLO secondary lymphoid organs; *TLS* tertiary lymphoid structures; *MLC* memory lymphocyte clusters; *HEV* high endothelial venules; T_{CM} central memory T cells; T_{RM} tissue-resident memory T cells; *FDC* follicular dendritic cells; *LT* lymphotoxin; *CCL* CC-chemokine ligand; *CXCL* CXC-chemokine ligand

5.1 MLC Formation Following HSV-2 Infection

HSV-2 is one of the most prevalent sexually transmitted pathogens with more than 400 million people infected worldwide. HSV-2 is a double-stranded DNA virus that belongs to the *Herpesviridae* family and is a major cause of genital herpes, symptoms of which include recurrent local inflammation along with severe pain in genital tissues leading to reduced quality of life (Schiffer et al. 2018). Following massive HSV-2 replication in the genital epithelium, the virus enters peripheral nerve terminals and then travels to dorsal root ganglia to replicate and/or lead to latency (Diefenbach et al. 2008). During latent infection in neuronal tissues, viral particles and viral DNA are rarely detected in vaginal and neuronal tissues. Instead, latency associated transcripts are abundantly localized in neuronal cell bodies but not genital tissues. Despite the lack of HSV-2 replication in vaginal tissues, lymphoid aggregates have been observed in human and mice (Gillgrass et al. 2005; Iijima et al. 2008a; Kiviat et al. 1990; Zhu et al. 2009), although their significance remained unclear until recently. To prevent HSV-2 infection in genital tissues and to treat HSV-2-mediated diseases, a large number of prophylactic vaccine trials have been performed; however, none of these trials has succeeded in generating a genital herpes vaccine (Awasthi et al. 2014; Belshe et al. 2012; Corey et al. 1999; Stanberry et al. 2002). In a murine model of genital herpes infection, intravaginal immunization with live attenuated HSV-2 was highly effective at inducing protective immune responses and the protection against HSV-2 (McDermott et al. 1984; Parr et al. 1994). Furthermore, at least four weeks after thymidine kinase negative (TK^-) HSV-2 intravaginal immunization, both T cells and IFN-γR signaling were required to inhibit virus replication in murine vaginal tissues following wild-type (WT) HSV-2 intravaginal challenge (Milligan and Bernstein 1997;

Milligan et al. 1998; Parr and Parr 1998, 1999). Likewise, in a guinea pig model of vaginal HSV-2 infection, $CD4^+$ T cells were required to the control HSV recurrence (Bourne et al. 2019). In contrast, intravaginal immunization with TK^- HSV-2 in B cell-deficient mice protected against HSV-2 secondary challenge, whereas intranasal immunization with attenuated virus in B cell-deficient mice allowed WT HSV-2 to invade neuronal tissues to replicate following intravaginal challenge (Iijima and Iwasaki 2016; Milligan et al. 2004). Furthermore, HSV-2-specific Ab required circulating $CD4^+$ T_M cells to achieve protection in neuronal tissues against WT HSV-2 intravaginal challenge (Iijima and Iwasaki 2016; Morrison et al. 2001). These findings suggested that tissue-resident immunity in vaginal tissues is critical for rapid viral clearance to block viral spreading into neuronal tissues. Certainly, HSV-specific $CD4^+$ T cells are found to accumulate in lamina propria of vaginal tissues four weeks after intravaginal immunization with TK^- HSV-2 (Iijima et al. 2008a). Furthermore, $CD4^+$ T cells form clusters beneath the vaginal epithelium (Roth et al. 2013). To inhibit HSV-2 replication in vaginal tissues, MHC class II^+ APC, including $CD11c^+$ DC and $CD20^+$ B cells are required for the IFN-γ-mediated but not cytotoxic function of $CD4^+$ T_M following HSV-2 secondary challenge (Iijima et al. 2008a). To dissect the mechanism for the maintenance of $CD4^+$ T_M in vaginal tissues following TK^- HSV-2 immunization, immunized C57BL/6 mice were conjoined with immunized congenic C57BL/6 mice to examine whether vaginal $CD4^+$ T_M constitute a tissue-resident population or a continuously circulating population. Two to seven weeks after surgery, HSV-2-specific $CD4^+$ T_M were predominantly a host-derived and not a blood-derived population (Iijima and Iwasaki 2014), indicating that these $CD4^+$ T_M are identified as T_{RM}. Furthermore, HSV-2-specific $CD4^+$ T_M bearing TCRVβ1 accumulate within MLC, although viral antigen sequence recognized by TCRVβ1 remains unclear. Regarding phenotypic features of $CD4^+$ T_{RM} in MLC, CD44, CD69, and CD49d are highly expressed, while CD103 and CD62L are rarely detected in HSV-2-specific $CD4^+$ T_{RM}. Remarkably, the expression of KLF2, KLF13, CCL5, and S1PR1 is barely observed in $CD4^+$ T_{RM} in MLC. In contrast, $CD4^+$ T_{RM} in MLC express T-bet, Bcl-xL and Bcl-2, indicating that IFN-γ-producing $CD4^+$ T_{RM} in MLC are a long-lived population (Iijima and Iwasaki 2014) (unpublished data). To maintain the formation of MLC, $CD11b^+$ macrophages within MLC are required and these cells secrete CCL5. Although viral genomic DNA or RNA transcripts have not been detected in vaginal tissues of TK^- HSV-2 immunized mice, constitutive low level secretion of IFN-γ from $CD4^+$ T_{RM} is observed in vaginal tissues (Iijima and Iwasaki 2014), indicating that $CD11b^+$ macrophages that stimulate $CD4^+$ T_{RM} to secrete IFN-γ produce CCL5 as a feedback loop mechanism. One of the mechanisms regarding MLC maintenance has, therefore, been elucidated but adhesion molecules and transcription factors that define $CD4^+$ T_{RM} and APC within MLC remain unknown. In the lamina propria of vaginal tissues, collagen type III, collagen type IV or desmin but not collagen type I is broadly distributed as major constituents of basement membranes, connective tissues and filaments (Fig. 2). In addition, PDPN/$gp38^+$ fibroblastic reticular cells and ER-$TR7^+$ fibroblasts are not found within MLC (Fig. 2). Future studies may

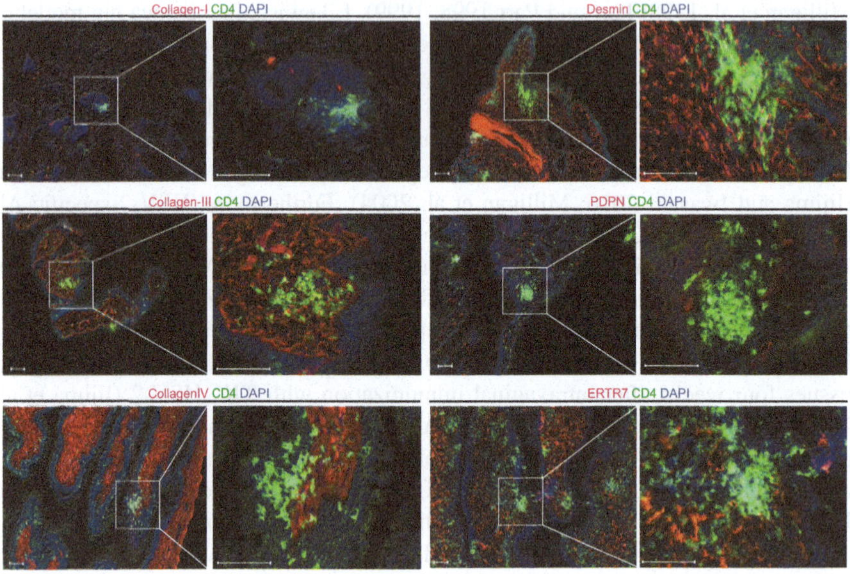

Fig. 2 Expression of collagen I, collagen III, collagen IV, desmin, PDPN or ER-TR7 in the lamina propria of vaginal tissues following immunization with attenuated HSV-2. C57BL/6 mice were immunized intravaginally with attenuated HSV-2. Five weeks later, frozen sections of vaginal tissue were stained with antibodies against collagen I, collagen III, collagen IV, desmin, PDPN/gp38 or ER-TR7 (red) and CD4 (green). Nuclei were stained by 4′,6′-diamidino-2-phenylindole (DAPI) (blue). Images were captured using a 10x or 40x objective lens. Scale bars indicate 100 μm

elucidate in more detail the interaction between $CD4^+$ T_{RM} and the network of fibroblastic reticular cells in genital tissues.

5.2 MLC Formation Following Chlamydia *Infection*

Chlamydia trachomatis infection of the genital mucosa causes an STD that is prevalent worldwide. Following invasion into the columnar epithelial cell lining of the endocervix and endometrial epithelium of the fallopian tubes, *Chlamydia trachomatis* triggers persistent inflammation leading to cause urethritis, proctitis, cervicitis, endometritis, and salpingitis (Poston and Darville 2018). More than six distinct types of curative antimicrobial drugs are available; however, there is no effective vaccine to prevent *Chlamydia* infection because of several strategies to evade the host immune response. Protective immunity against *Chlamydia* infection develops in DLN and genital tissues (Johnson and Brunham 2016); however, there is currently no critical target to exploit for development of an effective vaccine. Several groups has demonstrated that mice deficient in CD4, MHC class II, IL-12 or IFN-γ have increased susceptibility to genital *Chlamydia* infection (Morrison et al.

1995; Morrison et al. 2000; Perry et al. 1997; Wang et al. 1999), indicating that IFN-γ producing $CD4^+$ T cells are required for the inhibition of *Chlamydia* replication. Similarly to genital HSV-2 infection, lymphoid aggregates containing $CD4^+$ T cells beneath the epithelium have been observed after genital *Chlamydia* infection (Kiviat et al. 1990; Morrison and Morrison 2000). Interestingly, in children with active trachoma caused by ocular *Chlamydia* infection, lymphoid follicles also develop in the conjunctiva, in which the majority of immune cells are B cells not T cells (el-Asrar et al. 1989). In contrast, in adults infected with *Chlamydia trachomatis*, along with conjunctival scarring, lymphoid aggregates were observed in conjunctiva and a large number of $CD4^+$ T cells but not B cells formed clusters (Reacher et al. 1991), indicating that age-related factors might affect the composition of immune cells in lymphoid aggregates. It remains unknown whether age-related factors also influence the generation of MLC following genital *Chlamydia* infection. In a murine model of genital *Chlamydia* infection, $CD11b^+$ cells in addition to $CD4^+$ cells are the predominant cell populations within MLC-like structures in uterine tissues, which are analogous to MLC in the conjunctiva of adults (Morrison and Morrison 2000). Although the existence of MLC following *Chlamydia* infection has been confirmed by several groups (Johnson and Brunham 2016), it remains totally unknown whether $CD4^+$ T_M within MLC are a tissue-resident population or if they maintain an effector function for some time after *Chlamydia* infection. Stary G et al. (2015) elegantly demonstrated that intrauterine and intranasal immunization with charge-switching synthetic adjuvant particles containing UV-inactivated *Chlamydia trachomatis* succeeded in protecting against a genital *Chlamydia* challenge by generating $CD4^+$ T_{RM} in the uterine mucosa (Stary et al. 2015). Interestingly, immunization with UV-inactivated *Chlamydia* through the intrauterine route failed to achieve the protection against reinfection with *Chlamydia* because of massive accumulation of $FoxP3^+$ regulatory T cells in the uterine mucosa through the activation of tolerogenic $CD103^+$ DC. This study sheds light on the future development of an effective mucosal vaccine against *Chlamydia* infection based on the generation of protective T_{RM} in mucosal tissues.

6 Functional Features of MLC Against Sexually Transmitted Pathogens

There is currently no way to block the infection of genital epithelia by sexually transmitted pathogens, except for two viruses, HPV and HBV. In the case of HSV-2 infection, HSV gD initially binds to nectin-1 expressed on vaginal epithelial cells (Linehan et al. 2004). This interaction results in a conformational change in HSV gD so that the gH/gL heterodimer with gD induces a conformational change in HSV gB. Finally, HSV is able to fuse with nectin-1^+ epithelia through HSV gB to deliver the virion capsids containing HSV DNA (Eisenberg et al. 2012). These glycoproteins are potential targets for prophylactic and therapeutic vaccines to induce durable memory responses of T cells and B cells. In a mouse model of genital herpes, intravaginal immunization with TK^- HSV-2 establishes complete protection against vaginal HSV-2 challenge (Parr et al. 1994). In contrast, although large numbers of T cells from TK^- HSV-2 immunized mice, which include HSV-2-specific T_E from DLN, were adoptively transferred, the recipients were only partially protected against genital herpes infection (McDermott et al. 1989), indicating that the generation of genital immunity following intravaginal immunization with TK^- HSV-2 is required for the establishment of complete protection. Subsequently, intravaginal immunization with TK^- HSV-2 was found to maximize the generation of a tissue-resident population in vaginal tissues, including MLC formation for the retention of HSV-2-specific $CD4^+$ T_{RM} (Iijima and Iwasaki 2014). To directly elucidate the requirement of tissue-resident immunity, TK^- HSV-2 immunized mice were conjoined with naïve mice to share blood circulation. In naïve pairs, HSV-2-specific cell populations, including memory T and B cells and Ab, were observed in blood, while T_{RM} were not established in vaginal tissues. Following WT HSV-2 challenge of the naïve pair, viral clearance was significantly delayed compared with challenge of the immune pair, indicating that T_{RM}, especially $CD4^+$ T_{RM}, are required for rapid viral clearance (Iijima and Iwasaki 2014).

The majority of $CD4^+$ T_{RM} are localized in the lamina propria of mucosal tissues or dermal tissues of the skin (Collins et al. 2016; Iijima and Iwasaki 2014), indicating that $CD4^+$ T_{RM} also play a prominent role in the front-line of defense against invading pathogens. Following re-encounter with the same pathogen, $CD4^+$ T_{RM} secrete high levels of IFN-γ within 12 h (Glennie et al. 2015; Iijima and Iwasaki 2014; Oh et al. 2019), suggesting that local antigen-presenting cells present viral antigens to $CD4^+$ T_{RM} instead of migrating into DLN. In addition, $CD4^+$ T_{RM} also trigger accumulation of monocytes, NK cells, and B cells from blood circulation and induce local dendritic cell activation through IFN-γ secretion and cytotoxic activity (Beura et al. 2019). Furthermore, disruption of MLC formation by $CD11b^+$ cell depletion reduced the level of CXCL9 expression in vaginal tissues following WT HSV-2 challenge, causing failure of memory B cells to migrate into vaginal tissues to secrete HSV-2-specific IgGs (Oh et al. 2019). This indicates that MLC formation is a platform for the maintenance of T_{RM} to rapidly exert their effector functions upon reencountering of invading pathogens.

7 Summary and Outlook

The generation of a robust immune response in peripheral tissues has a substantial impact on the protection against invading pathogens, the onset of autoimmune diseases, and tumor progression. On the one hand, SLO including LN are pre-programmed encapsulated organs that organize both innate and adaptive immune responses to maximize the initiation of antigen-specific immunity. On the other hand, TLS formation represents a non-encapsulated congregation of lymphocytes and stromal cells in peripheral tissues. In both SLO and TLS, stromal cell-derived HEV develop to recruit CD62L^{+} naïve B and T cells and T_{CM} so that these lymphocytes can undergo transition from blood circulation to B cell zones or T cell zones (Fig. 1a and b) (Table 1). In contrast, MLC lack HEV and B cell areas, and naïve T and B cells do not reside in MLC (Fig. 1c). Instead, MLC consist of CD11b^{+} APC with CD44^{+} CD62L^{-} CD69^{+} CD4^{+} T_{RM} and CD8^{+} T_{RM} beneath mucosal epithelial layers. SLO and TLS are, therefore, suggested to be inductive sites for antigen-specific immune responses upon encountering newly delivered antigens, while MLC appear to be specialized in the maintenance of tissue-resident memory populations in peripheral tissues to initiate rapid immune response at the mucosal surface. Currently, the molecular mechanisms underlying the generation of MLC are poorly understood. The cellular composition and distribution of MLC are obviously distinct from those of SLO and TLS (Table 1). In association with the deployment of immune cells in MLC, the findings regarding the neogenesis of iBALT formation within or outside of B cell areas following repetitive inhalation of heat-killed *P. aeruginosa* is significantly informative (Fleige et al. 2014). This study demonstrated that BALT formation mainly lacks B cell areas containing CXCL12^{+} stromal cells in IL-17A and F deficient mice following inoculation of heat-inactivated *P. aeruginosa*, the formation consisting entirely of CD3^{+} T cells. Consequently, the BALT formation appears to resemble MLC formation, although it remains unclear whether the B cell-deficient BALT contain PNAd^{+} HEV and naïve T cells.

The following questions concerning MLC formation remain to be answered.

1. What type of cells initiates MLC formation in genital tissue?
2. What cytokines or chemokines are responsible for the generation of MLC?
3. Which integrins and adhesion molecules expressed on T_{RM} are involved in the interaction with stromal cells?
4. How do CD4^{+} T_{RM} in the lamina propria attach to epithelial layers?
5. How is the effector function of T_{RM} in MLC maintained?
6. What type of APC is responsible for reactivation of T_{RM} in MLC?
7. How do antigen-captured APC reactivate T_{RM} in MLC to exert effector function?
8. How do reactivated T_{RM} in MLC migrate to the site of viral/bacterial replication?

Acknowledgements We thank A. Iwasaki, S. Shimada and J. Allen for critical reading of this manuscript. We apologize to the many scientists whose work we were not able to mention due to space limitations. This work was supported by the Japan Agency for Medical Research and Development (AMED) (17fm0208011h0001, 18fm0208011h0002, 19fm0208011h0003), KAKENHI (16H07504, 18H02857, 19K22645), the Takeda Science Foundation, the Mochida Memorial Foundation for Medical and Pharmaceutical Research, the Daiichi Sankyo Foundation of Life Science, the Astellas Foundation for Research on Metabolic Disorders, the Uehara Memorial Foundation, Bristol-Myers Squibb Foundation Grants and Novartis Pharma Research Grants.

References

Acosta-Ramirez E, Tram C, Kampen RM, Tillman MR, Schwendener RA, Xing Z, Halperin SA, Wang J (2016) Respiratory macrophages regulate CD4 T memory responses to mucosal immunization with recombinant adenovirus-based vaccines. Cell Immunol 310:53–62. https://doi.org/10.1016/j.cellimm.2016.07.006

Adachi T, Kobayashi T, Sugihara E, Yamada T, Ikuta K, Pittaluga S, Saya H, Amagai M, Nagao K (2015) Hair follicle-derived IL-7 and IL-15 mediate skin-resident memory T cell homeostasis and lymphoma. Nat Med 21(11):1272–1279. https://doi.org/10.1038/nm.3962

Ager A (2017) High endothelial venules and other blood vessels: critical regulators of lymphoid organ development and function. Front Immunol 8:45. https://doi.org/10.3389/fimmu.2017.00045

Alsughayyir J, Pettigrew GJ, Motallebzadeh R (2017) Spoiling for a fight: B lymphocytes as initiator and effector populations within tertiary lymphoid organs in autoimmunity and transplantation. Front Immunol 8:1639. https://doi.org/10.3389/fimmu.2017.01639

Amezcua Vesely MC, Pallis P, Bielecki P, Low JS, Zhao J, Harman CCD, Kroehling L, Jackson R, Bailis W, Licona-Limon P, Xu H, Iijima N, Pillai PS, Kaplan DH, Weaver CT, Kluger Y, Kowalczyk MS, Iwasaki A, Pereira JP, Esplugues E, Gagliani N, Flavell RA (2019) Effector TH17 cells give rise to long-lived TRM cells that are essential for an immediate response against bacterial infection. Cell 178(5):1176–1188 e1115. https://doi.org/10.1016/j.cell.2019.07.032

Anipindi VC, Bagri P, Roth K, Dizzell SE, Nguyen PV, Shaler CR, Chu DK, Jimenez-Saiz R, Liang H, Swift S, Nazli A, Kafka JK, Bramson J, Xing Z, Jordana M, Wan Y, Snider DP, Stampfli MR, Kaushic C (2016) Estradiol enhances CD4 + T-cell anti-viral immunity by priming vaginal DCs to induce Th17 responses via an IL-1-dependent pathway. PLoS Pathog 12(5):e1005589. https://doi.org/10.1371/journal.ppat.1005589

Awasthi S, Shaw C, Friedman H (2014) Improving immunogenicity and efficacy of vaccines for genital herpes containing herpes simplex virus glycoprotein D. Expert Rev Vaccines 13 (12):1475–1488. https://doi.org/10.1586/14760584.2014.951336

Badovinac VP, Porter BB, Harty JT (2002) Programmed contraction of CD8(+) T cells after infection. Nat Immunol 3(7):619–626. https://doi.org/10.1038/ni804

Ballweber L, Robinson B, Kreger A, Fialkow M, Lentz G, McElrath MJ, Hladik F (2011) Vaginal langerhans cells nonproductively transporting HIV-1 mediate infection of T cells. J Virol 85 (24):13443–13447. https://doi.org/10.1128/JVI.05615-11

Barone F, Gardner DH, Nayar S, Steinthal N, Buckley CD, Luther SA (2016) Stromal fibroblasts in tertiary lymphoid structures: a novel target in chronic inflammation. Front Immunol 7:477. https://doi.org/10.3389/fimmu.2016.00477

Beaman KD, Jaiswal MK, Dambaeva S, Gilman-Sachs A (2014) Future directions of clinical laboratory evaluation of pregnancy. Cell Mol Immunol 11(6):582–588. https://doi.org/10.1038/cmi.2014.62

Behr FM, Kragten NAM, Wesselink TH, Nota B, van Lier RAW, Amsen D, Stark R, Hombrink P, van Gisbergen K (2019) Blimp-1 rather than hobit drives the formation of tissue-resident memory CD8(+) T cells in the lungs. Front Immunol 10:400. https://doi.org/10.3389/fimmu.2019.00400

Belshe RB, Leone PA, Bernstein DI, Wald A, Levin MJ, Stapleton JT, Gorfinkel I, Morrow RL, Ewell MG, Stokes-Riner A, Dubin G, Heineman TC, Schulte JM, Deal CD (2012) Herpevac trial for woman. Efficacy results of a trial of a herpes simplex vaccine. N Engl J Med 366 (1):34–43. https://doi.org/10.1056/NEJMoa1103151

Bergsbaken T, Bevan MJ (2015) Proinflammatory microenvironments within the intestine regulate the differentiation of tissue-resident CD8(+) T cells responding to infection. Nat Immunol 16 (4):406–414. https://doi.org/10.1038/ni.3108

Beura LK, Fares-Frederickson NJ, Steinert EM, Scott MC, Thompson EA, Fraser KA, Schenkel JM, Vezys V, Masopust D (2019) CD4(+) resident memory T cells dominate immunosurveillance and orchestrate local recall responses. J Exp Med 216(5):1214–1229. https://doi.org/10.1084/jem.20181365

Beura LK, Wijeyesinghe S, Thompson EA, Macchietto MG, Rosato PC, Pierson MJ, Schenkel JM, Mitchell JS, Vezys V, Fife BT, Shen S, Masopust D (2018) T Cells in nonlymphoid tissues give rise to lymph-node-resident memory t cells. Immunity 48(2):327–338 e325. https://doi.org/10.1016/j.immuni.2018.01.015

Bistrup A, Tsay D, Shenoy P, Singer MS, Bangia N, Luther SA, Cyster JG, Ruddle NH, Rosen SD (2004) Detection of a sulfotransferase (HEC-GlcNAc6ST) in high endothelial venules of lymph nodes and in high endothelial venule-like vessels within ectopic lymphoid aggregates: relationship to the MECA-79 epitope. Am J Pathol 164(5):1635–1644. https://doi.org/10.1016/S0002-9440(10)63722-4

Borges da Silva H, Beura LK, Wang H, Hanse EA, Gore R, Scott MC, Walsh DA, Block KE, Fonseca R, Yan Y, Hippen KL, Blazar BR, Masopust D, Kelekar A, Vulchanova L, Hogquist KA, Jameson SC (2018) The purinergic receptor P2RX7 directs metabolic fitness of long-lived memory CD8(+) T cells. Nature 559(7713):264–268. https://doi.org/10.1038/s41586-018-0282-0

Bottcher JP, Beyer M, Meissner F, Abdullah Z, Sander J, Hochst B, Eickhoff S, Rieckmann JC, Russo C, Bauer T, Flecken T, Giesen D, Engel D, Jung S, Busch DH, Protzer U, Thimme R, Mann M, Kurts C, Schultze JL, Kastenmuller W, Knolle PA (2015) Functional classification of memory CD8(+) T cells by CX3CR1 expression. Nat Commun 6:8306. https://doi.org/10.1038/ncomms9306

Bourne N, Perry CL, Banasik BN, Miller AL, White M, Pyles RB, Schafer H, Milligan GN (2019) Increased frequency of virus shedding by herpes simplex virus 2-infected guinea pigs in the absence of CD4(+) T lymphocytes. J Virol 93(4). https://doi.org/10.1128/jvi.01721-18

Bromley SK, Mempel TR, Luster AD (2008) Orchestrating the orchestrators: chemokines in control of T cell traffic. Nat Immunol 9(9):970–980. https://doi.org/10.1038/ni.f.213

Bromley SK, Thomas SY, Luster AD (2005) Chemokine receptor CCR7 guides T cell exit from peripheral tissues and entry into afferent lymphatics. Nat Immunol 6(9):895–901. https://doi.org/10.1038/ni1240

Capucha T, Mizraji G, Segev H, Blecher-Gonen R, Winter D, Khalaileh A, Tabib Y, Attal T, Nassar M, Zelentsova K, Kisos H, Zenke M, Sere K, Hieronymus T, Burstyn-Cohen T, Amit I, Wilensky A, Hovav AH (2015) Distinct murine mucosal langerhans cell subsets develop from pre-dendritic cells and monocytes. Immunity 43(2):369–381. https://doi.org/10.1016/j.immuni.2015.06.017

Casey KA, Fraser KA, Schenkel JM, Moran A, Abt MC, Beura LK, Lucas PJ, Artis D, Wherry EJ, Hogquist K, Vezys V, Masopust D (2012) Antigen-independent differentiation and maintenance of effector-like resident memory T cells in tissues. J Immunol 188(10):4866–4875. https://doi.org/10.4049/jimmunol.1200402

Chang JE, Turley SJ (2015) Stromal infrastructure of the lymph node and coordination of immunity. Trends Immunol 36(1):30–39. https://doi.org/10.1016/j.it.2014.11.003

Chen C, Song X, Wei W, Zhong H, Dai J, Lan Z, Li F, Yu X, Feng Q, Wang Z, Xie H, Chen X, Zeng C, Wen B, Zeng L, Du H, Tang H, Xu C, Xia Y, Xia H, Yang H, Wang J, Wang J, Madsen L, Brix S, Kristiansen K, Xu X, Li J, Wu R, Jia H (2017) The microbiota continuum along the female reproductive tract and its relation to uterine-related diseases. Nat Commun 8(1):875. https://doi.org/10.1038/s41467-017-00901-0

Chorro L, Sarde A, Li M, Woollard KJ, Chambon P, Malissen B, Kissenpfennig A, Barbaroux JB, Groves R, Geissmann F (2009) Langerhans cell (LC) proliferation mediates neonatal development, homeostasis, and inflammation-associated expansion of the epidermal LC network. J Exp Med 206(13):3089–3100. https://doi.org/10.1084/jem.20091586

Clark RA (2015) Resident memory T cells in human health and disease. Sci Transl Med 7(269):269rv261. https://doi.org/10.1126/scitranslmed.3010641

Clark RA, Watanabe R, Teague JE, Schlapbach C, Tawa MC, Adams N, Dorosario AA, Chaney KS, Cutler CS, Leboeuf NR, Carter JB, Fisher DC, Kupper TS (2012) Skin effector memory T cells do not recirculate and provide immune protection in alemtuzumab-treated

CTCL patients. Sci Transl Med 4(117):117ra117. https://doi.org/10.1126/scitranslmed.3003008

Collins MK, Tay CS, Erlebacher A (2009) Dendritic cell entrapment within the pregnant uterus inhibits immune surveillance of the maternal/fetal interface in mice. J Clin Invest 119(7):2062–2073. https://doi.org/10.1172/JCI38714

Collins N, Jiang X, Zaid A, Macleod BL, Li J, Park CO, Haque A, Bedoui S, Heath WR, Mueller SN, Kupper TS, Gebhardt T, Carbone FR (2016) Skin CD4(+) memory T cells exhibit combined cluster-mediated retention and equilibration with the circulation. Nat Commun 7:11514. https://doi.org/10.1038/ncomms11514

Corey L, Langenberg AG, Ashley R, Sekulovich RE, Izu AE, Douglas JM Jr, Handsfield HH, Warren T, Marr L, Tyring S, DiCarlo R, Adimora AA, Leone P, Dekker CL, Burke RL, Leong WP, Straus SE (1999) Recombinant glycoprotein vaccine for the prevention of genital HSV-2 infection: two randomized controlled trials. Chiron HSV Vaccine Study Group. JAMA 282(4):331–340. https://doi.org/10.1001/jama.282.4.331

Corsiero E, Delvecchio FR, Bombardieri M, Pitzalis C (2019) B cells in the formation of tertiary lymphoid organs in autoimmunity, transplantation and tumorigenesis. Curr Opin Immunol 57:46–52. https://doi.org/10.1016/j.coi.2019.01.004

Cuburu N, Kim R, Guittard GC, Thompson CD, Day PM, Hamm DE, Pang YS, Graham BS, Lowy DR, Schiller JT (2019) A prime-pull-amplify vaccination strategy to maximize induction of circulating and genital-resident intraepithelial CD8(+) memory T cells. J Immunol 202 (4):1250–1264. https://doi.org/10.4049/jimmunol.1800219

Debes GF, Arnold CN, Young AJ, Krautwald S, Lipp M, Hay JB, Butcher EC (2005) Chemokine receptor CCR7 required for T lymphocyte exit from peripheral tissues. Nat Immunol 6(9):889–894. https://doi.org/10.1038/ni1238

Denton AE, Carr EJ, Magiera LP, Watts AJB, Fearon DT (2019) Embryonic FAP(+) lymphoid tissue organizer cells generate the reticular network of adult lymph nodes. J Exp Med. https://doi.org/10.1084/jem.20181705

Deruaz M, Luster AD (2015) Chemokine-mediated immune responses in the female genital tract mucosa. Immunol Cell Biol 93(4):347–354. https://doi.org/10.1038/icb.2015.20

Diefenbach RJ, Miranda-Saksena M, Douglas MW, Cunningham AL (2008) Transport and egress of herpes simplex virus in neurons. Rev Med Virol 18(1):35–51. https://doi.org/10.1002/rmv.560

Drayton DL, Liao S, Mounzer RH, Ruddle NH (2006) Lymphoid organ development: from ontogeny to neogenesis. Nat Immunol 7(4):344–353. https://doi.org/10.1038/ni1330

Duluc D, Gannevat J, Anguiano E, Zurawski S, Carley M, Boreham M, Stecher J, Dullaers M, Banchereau J, Oh S (2013) Functional diversity of human vaginal APC subsets in directing T-cell responses. Mucosal Immunol 6(3):626–638. https://doi.org/10.1038/mi.2012.104

Egawa N, Egawa K, Griffin H, Doorbar J (2015) Human Papillomaviruses; epithelial tropisms, and the development of neoplasia. Viruses 7(7):3863–3890. https://doi.org/10.3390/v7072802

Eisenberg RJ, Atanasiu D, Cairns TM, Gallagher JR, Krummenacher C, Cohen GH (2012) Herpes virus fusion and entry: a story with many characters. Viruses 4(5):800–832. https://doi.org/10.3390/v4050800

el-Asrar AM, Van den Oord JJ, Geboes K, Missotten L, Emarah MH, Desmet V (1989) Immunopathology of trachomatous conjunctivitis. Br J Ophthalmol 73(4):276–282. https://doi.org/10.1136/bjo.73.4.276

Fleige H, Ravens S, Moschovakis GL, Bolter J, Willenzon S, Sutter G, Haussler S, Kalinke U, Prinz I, Forster R (2014) IL-17-induced CXCL12 recruits B cells and induces follicle formation in BALT in the absence of differentiated FDCs. J Exp Med 211(4):643–651. https://doi.org/10.1084/jem.20131737

Fridman WH, Zitvogel L, Sautes-Fridman C, Kroemer G (2017) The immune contexture in cancer prognosis and treatment. Nat Rev Clin Oncol 14(12):717–734. https://doi.org/10.1038/nrclinonc.2017.101

Furtado GC, Marinkovic T, Martin AP, Garin A, Hoch B, Hubner W, Chen BK, Genden E, Skobe M, Lira SA (2007) Lymphotoxin beta receptor signaling is required for inflammatory lymphangiogenesis in the thyroid. Proc Natl Acad Sci U S A 104(12):5026–5031. https://doi.org/10.1073/pnas.0606697104

Gerlach C, Moseman EA, Loughhead SM, Alvarez D, Zwijnenburg AJ, Waanders L, Garg R, de la Torre JC, von Andrian UH (2016) The chemokine receptor CX3CR1 defines three antigen-experienced CD8 T cell subsets with distinct roles in immune surveillance and homeostasis. Immunity 45(6):1270–1284. https://doi.org/10.1016/j.immuni.2016.10.018

Gillgrass AE, Tang VA, Towarnicki KM, Rosenthal KL, Kaushic C (2005) Protection against genital herpes infection in mice immunized under different hormonal conditions correlates with induction of vagina-associated lymphoid tissue. J Virol 79(5):3117–3126. https://doi.org/10.1128/JVI.79.5.3117-3126.2005

Glennie ND, Yeramilli VA, Beiting DP, Volk SW, Weaver CT, Scott P (2015) Skin-resident memory CD4 + T cells enhance protection against Leishmania major infection. J Exp Med 212 (9):1405–1414. https://doi.org/10.1084/jem.20142101

Gopinath S, Kim MV, Rakib T, Wong PW, van Zandt M, Barry NA, Kaisho T, Goodman AL, Iwasaki A (2018) Topical application of aminoglycoside antibiotics enhances host resistance to viral infections in a microbiota-independent manner. Nat Microbiol 3(5):611–621. https://doi.org/10.1038/s41564-018-0138-2

Guedj K, Khallou-Laschet J, Clement M, Morvan M, Gaston AT, Fornasa G, Dai J, Gervais-Taurel M, Eberl G, Michel JB, Caligiuri G, Nicoletti A (2014) M1 macrophages act as LTbetaR-independent lymphoid tissue inducer cells during atherosclerosis-related lymphoid neogenesis. Cardiovasc Res 101(3):434–443. https://doi.org/10.1093/cvr/cvt263

Hattori T, Chauhan SK, Lee H, Ueno H, Dana R, Kaplan DH, Saban DR (2011) Characterization of Langerin-expressing dendritic cell subsets in the normal cornea. Invest Ophthalmol Vis Sci 52(7):4598–4604. https://doi.org/10.1167/iovs.10-6741

Hegazy AN, Peine M, Helmstetter C, Panse I, Frohlich A, Bergthaler A, Flatz L, Pinschewer DD, Radbruch A, Lohning M (2010) Interferons direct Th2 cell reprogramming to generate a stable GATA-3(+)T-bet(+) cell subset with combined Th2 and Th1 cell functions. Immunity 32 (1):116–128. https://doi.org/10.1016/j.immuni.2009.12.004

Hervouet C, Luci C, Rol N, Rousseau D, Kissenpfennig A, Malissen B, Czerkinsky C, Anjuere F (2010) Langerhans cells prime IL-17-producing T cells and dampen genital cytotoxic responses following mucosal immunization. J Immunol 184(9):4842–4851. https://doi.org/10.4049/jimmunol.0901695

Hirai T, Zenke Y, Yang Y, Bartholin L, Beura LK, Masopust D, Kaplan DH (2019) Keratinocyte-mediated activation of the cytokine TGF-beta maintains skin recirculating memory CD8(+) T Cells. Immunity 50(5):1249–1261 e1245. https://doi.org/10.1016/j.immuni.2019.03.002

Hoeffel G, Wang Y, Greter M, See P, Teo P, Malleret B, Leboeuf M, Low D, Oller G, Almeida F, Choy SH, Grisotto M, Renia L, Conway SJ, Stanley ER, Chan JK, Ng LG, Samokhvalov IM, Merad M, Ginhoux F (2012) Adult Langerhans cells derive predominantly from embryonic fetal liver monocytes with a minor contribution of yolk sac-derived macrophages. J Exp Med 209(6):1167–1181. https://doi.org/10.1084/jem.20120340

Hogan RJ, Zhong W, Usherwood EJ, Cookenham T, Roberts AD, Woodland DL (2001) Protection from respiratory virus infections can be mediated by antigen-specific CD4(+) T cells that persist in the lungs. J Exp Med 193(8):981–986. https://doi.org/10.1084/jem.193.8.981

Hondowicz BD, An D, Schenkel JM, Kim KS, Steach HR, Krishnamurty AT, Keitany GJ, Garza EN, Fraser KA, Moon JJ, Altemeier WA, Masopust D, Pepper M (2016) Interleukin-2-dependent allergen-specific tissue-resident memory cells drive asthma. Immunity 44(1):155–166. https://doi.org/10.1016/j.immuni.2015.11.004

Hondowicz BD, Kim KS, Ruterbusch MJ, Keitany GJ, Pepper M (2018) IL-2 is required for the generation of viral-specific CD4(+) Th1 tissue-resident memory cells and B cells are essential for maintenance in the lung. Eur J Immunol 48(1):80–86. https://doi.org/10.1002/eji.201746928

Houser BL, Tilburgs T, Hill J, Nicotra ML, Strominger JL (2011) Two unique human decidual macrophage populations. J Immunol 186(4):2633–2642. https://doi.org/10.4049/jimmunol.1003153

Hovav AH (2018) Mucosal and skin langerhans cells—nurture calls. Trends Immunol 39(10):788–800. https://doi.org/10.1016/j.it.2018.08.007

Iijima N, Iwasaki A (2014) T cell memory. A local macrophage chemokine network sustains protective tissue-resident memory CD4 T cells. Science 346(6205):93–98. https://doi.org/10.1126/science.1257530

Iijima N, Iwasaki A (2015) Tissue instruction for migration and retention of TRM cells. Trends Immunol 36(9):556–564. https://doi.org/10.1016/j.it.2015.07.002

Iijima N, Iwasaki A (2016) Access of protective antiviral antibody to neuronal tissues requires CD4 T-cell help. Nature 533(7604):552–556. https://doi.org/10.1038/nature17979

Iijima N, Linehan MM, Saeland S, Iwasaki A (2007) Vaginal epithelial dendritic cells renew from bone marrow precursors. Proc Natl Acad Sci U S A 104(48):19061–19066. https://doi.org/10.1073/pnas.0707179104

Iijima N, Mattei LM, Iwasaki A (2011) Recruited inflammatory monocytes stimulate antiviral Th1 immunity in infected tissue. Proc Natl Acad Sci U S A 108(1):284–289. https://doi.org/10.1073/pnas.1005201108

Iijima N, Linehan MM, Zamora M, Butkus D, Dunn R, Kehry MR, Laufer TM, Iwasaki A (2008a) Dendritic cells and B cells maximize mucosal Th1 memory response to herpes simplex virus. J Exp Med 205(13):3041–3052. https://doi.org/10.1084/jem.20082039

Iijima N, Thompson JM, Iwasaki A (2008b) Dendritic cells and macrophages in the genitourinary tract. Mucosal Immunol 1(6):451–459. https://doi.org/10.1038/mi.2008.57

Iwasaki A (2007) Mucosal dendritic cells. Annu Rev Immunol 25:381–418. https://doi.org/10.1146/annurev.immunol.25.022106.141634

Iwasaki A (2010) Antiviral immune responses in the genital tract: clues for vaccines. Nat Rev Immunol 10(10):699–711. https://doi.org/10.1038/nri2836

Iwasaki A (2016) Exploiting mucosal immunity for antiviral vaccines. Annu Rev Immunol 34:575–608. https://doi.org/10.1146/annurev-immunol-032414-112315

Jameson SC, Masopust D (2018) Understanding subset diversity in T cell memory. Immunity 48(2):214–226. https://doi.org/10.1016/j.immuni.2018.02.010

Jiang X, Clark RA, Liu L, Wagers AJ, Fuhlbrigge RC, Kupper TS (2012) Skin infection generates non-migratory memory CD8 + T(RM) cells providing global skin immunity. Nature 483(7388):227–231. https://doi.org/10.1038/nature10851

Johansson EL, Rudin A, Wassen L, Holmgren J (1999) Distribution of lymphocytes and adhesion molecules in human cervix and vagina. Immunology 96(2):272–277. https://doi.org/10.1046/j.1365-2567.1999.00675.x

Johnson RM, Brunham RC (2016) Tissue-resident T cells as the central paradigm of chlamydia immunity. Infect Immun 84(4):868–873. https://doi.org/10.1128/IAI.01378-15

Joshi NS, Cui W, Chandele A, Lee HK, Urso DR, Hagman J, Gapin L, Kaech SM (2007) Inflammation directs memory precursor and short-lived effector CD8(+) T cell fates via the graded expression of T-bet transcription factor. Immunity 27(2):281–295. https://doi.org/10.1016/j.immuni.2007.07.010

Joura E, Bautista O, Luxembourg A (2015) A 9-valent HPV vaccine in women. N Engl J Med 372(26):2568–2569. https://doi.org/10.1056/NEJMc1504359

Kabashima K, Honda T, Ginhoux F, Egawa G (2019) The immunological anatomy of the skin. Nat Rev Immunol 19(1):19–30. https://doi.org/10.1038/s41577-018-0084-5

Kather A, Chantakru S, He H, Minhas K, Foster R, Markert UR, Pfeffer K, Croy BA (2003) Neither lymphotoxin alpha nor lymphotoxin beta receptor expression is required for biogenesis of lymphoid aggregates or differentiation of natural killer cells in the pregnant mouse uterus. Immunology 108(3):338–345. https://doi.org/10.1046/j.1365-2567.2003.01586.x

Kiviat NB, Paavonen JA, Wolner-Hanssen P, Critchlow CW, Stamm WE, Douglas J, Eschenbach DA, Corey LA, Holmes KK (1990) Histopathology of endocervical infection

caused by *Chlamydia trachomatis*, herpes simplex virus, *Trichomonas vaginalis*, and *Neisseria gonorrhoeae*. Hum Pathol 21(8):831–837. https://doi.org/10.1016/0046-8177(90)90052-7
Klicznik MM, Morawski PA, Hollbacher B, Varkhande SR, Motley SJ, Kuri-Cervantes L, Goodwin E, Rosenblum MD, Long SA, Brachtl G, Duhen T, Betts MR, Campbell DJ, Gratz IK (2019) Human CD4(+)CD103(+) cutaneous resident memory T cells are found in the circulation of healthy individuals. Sci Immunol 4(37). https://doi.org/10.1126/sciimmunol.aav8995
Klonowski KD, Williams KJ, Marzo AL, Blair DA, Lingenheld EG, Lefrancois L (2004) Dynamics of blood-borne CD8 memory T cell migration in vivo. Immunity 20(5):551–562
Koopman LA, Kopcow HD, Rybalov B, Boyson JE, Orange JS, Schatz F, Masch R, Lockwood CJ, Schachter AD, Park PJ, Strominger JL (2003) Human decidual natural killer cells are a unique NK cell subset with immunomodulatory potential. J Exp Med 198(8):1201–1212. https://doi.org/10.1084/jem.20030305
Ledgerwood LG, Lal G, Zhang N, Garin A, Esses SJ, Ginhoux F, Merad M, Peche H, Lira SA, Ding Y, Yang Y, He X, Schuchman EH, Allende ML, Ochando JC, Bromberg JS (2008) The sphingosine 1-phosphate receptor 1 causes tissue retention by inhibiting the entry of peripheral tissue T lymphocytes into afferent lymphatics. Nat Immunol 9(1):42–53. https://doi.org/10.1038/ni1534
Lee HK, Zamora M, Linehan MM, Iijima N, Gonzalez D, Haberman A, Iwasaki A (2009) Differential roles of migratory and resident DCs in T cell priming after mucosal or skin HSV-1 infection. J Exp Med 206(2):359–370. https://doi.org/10.1084/jem.20080601
Linehan MM, Richman S, Krummenacher C, Eisenberg RJ, Cohen GH, Iwasaki A (2004) In vivo role of nectin-1 in entry of herpes simplex virus type 1 (HSV-1) and HSV-2 through the vaginal mucosa. J Virol 78(5):2530–2536. https://doi.org/10.1128/jvi.78.5.2530-2536.2004
Lochner M, Ohnmacht C, Presley L, Bruhns P, Si-Tahar M, Sawa S, Eberl G (2011) Microbiota-induced tertiary lymphoid tissues aggravate inflammatory disease in the absence of RORgamma t and LTi cells. J Exp Med 208(1):125–134. https://doi.org/10.1084/jem.20100052
Lucchesi D, Pitzalis C, Bombardieri M (2014) EBV and other viruses as triggers of tertiary lymphoid structures in primary Sjogren's syndrome. Expert Rev Clin Immunol 10(4):445–455. https://doi.org/10.1586/1744666X.2014.892417
Lund JM, Hsing L, Pham TT, Rudensky AY (2008) Coordination of early protective immunity to viral infection by regulatory T cells. Science 320(5880):1220–1224. https://doi.org/10.1126/science.1155209
Mackay LK, Minnich M, Kragten NA, Liao Y, Nota B, Seillet C, Zaid A, Man K, Preston S, Freestone D, Braun A, Wynne-Jones E, Behr FM, Stark R, Pellicci DG, Godfrey DI, Belz GT, Pellegrini M, Gebhardt T, Busslinger M, Shi W, Carbone FR, van Lier RA, Kallies A, van Gisbergen KP (2016) Hobit and Blimp1 instruct a universal transcriptional program of tissue residency in lymphocytes. Science 352(6284):459–463. https://doi.org/10.1126/science.aad2035
Mackay LK, Rahimpour A, Ma JZ, Collins N, Stock AT, Hafon ML, Vega-Ramos J, Lauzurica P, Mueller SN, Stefanovic T, Tscharke DC, Heath WR, Inouye M, Carbone FR, Gebhardt T (2013) The developmental pathway for CD103(+)CD8 + tissue-resident memory T cells of skin. Nat Immunol 14(12):1294–1301. https://doi.org/10.1038/ni.2744
Mackay LK, Stock AT, Ma JZ, Jones CM, Kent SJ, Mueller SN, Heath WR, Carbone FR, Gebhardt T (2012) Long-lived epithelial immunity by tissue-resident memory T (TRM) cells in the absence of persisting local antigen presentation. Proc Natl Acad Sci U S A 109(18):7037–7042. https://doi.org/10.1073/pnas.1202288109
Mackay LK, Wynne-Jones E, Freestone D, Pellicci DG, Mielke LA, Newman DM, Braun A, Masson F, Kallies A, Belz GT, Carbone FR (2015) T-box transcription factors combine with the cytokines TGF-beta and IL-15 to control tissue-resident memory T cell fate. Immunity 43(6):1101–1111. https://doi.org/10.1016/j.immuni.2015.11.008
Marshall HD, Chandele A, Jung YW, Meng H, Poholek AC, Parish IA, Rutishauser R, Cui W, Kleinstein SH, Craft J, Kaech SM (2011) Differential expression of Ly6C and T-bet distinguish

effector and memory Th1 CD4(+) cell properties during viral infection. Immunity 35(4):633–646. https://doi.org/10.1016/j.immuni.2011.08.016
Masopust D, Soerens AG (2019) Tissue-resident T cells and other resident leukocytes. Annu Rev Immunol 37:521–546. https://doi.org/10.1146/annurev-immunol-042617-053214
Masopust D, Vezys V, Marzo AL, Lefrancois L (2001) Preferential localization of effector memory cells in nonlymphoid tissue. Science 291(5512):2413–2417. https://doi.org/10.1126/science.1058867
Matloubian M, Lo CG, Cinamon G, Lesneski MJ, Xu Y, Brinkmann V, Allende ML, Proia RL, Cyster JG (2004) Lymphocyte egress from thymus and peripheral lymphoid organs is dependent on S1P receptor 1. Nature 427(6972):355–360. https://doi.org/10.1038/nature02284
McCully ML, Ladell K, Hakobyan S, Mansel RE, Price DA, Moser B (2012) Epidermis instructs skin homing receptor expression in human T cells. Blood 120(23):4591–4598. https://doi.org/10.1182/blood-2012-05-433037
McDermott MR, Goldsmith CH, Rosenthal KL, Brais LJ (1989) T lymphocytes in genital lymph nodes protect mice from intravaginal infection with herpes simplex virus type 2. J Infect Dis 159(3):460–466. https://doi.org/10.1093/infdis/159.3.460
McDermott MR, Smiley JR, Leslie P, Brais J, Rudzroga HE, Bienenstock J (1984) Immunity in the female genital tract after intravaginal vaccination of mice with an attenuated strain of herpes simplex virus type 2. J Virol 51(3):747–753
Merad M, Manz MG, Karsunky H, Wagers A, Peters W, Charo I, Weissman IL, Cyster JG, Engleman EG (2002) Langerhans cells renew in the skin throughout life under steady-state conditions. Nat Immunol 3(12):1135–1141. https://doi.org/10.1038/ni852
Milligan GN, Bernstein DI (1997) Interferon-gamma enhances resolution of herpes simplex virus type 2 infection of the murine genital tract. Virology 229(1):259–268. https://doi.org/10.1006/viro.1997.8441
Milligan GN, Bernstein DI, Bourne N (1998) T lymphocytes are required for protection of the vaginal mucosae and sensory ganglia of immune mice against reinfection with herpes simplex virus type 2. J Immunol 160(12):6093–6100
Milligan GN, Dudley-McClain KL, Chu CF, Young CG (2004) Efficacy of genital T cell responses to herpes simplex virus type 2 resulting from immunization of the nasal mucosa. Virology 318 (2):507–515. https://doi.org/10.1016/j.virol.2003.10.010
Milner JJ, Toma C, Yu B, Zhang K, Omilusik K, Phan AT, Wang D, Getzler AJ, Nguyen T, Crotty S, Wang W, Pipkin ME, Goldrath AW (2017) Runx3 programs CD8(+) T cell residency in non-lymphoid tissues and tumours. Nature 552(7684):253–257. https://doi.org/10.1038/nature24993
Mohammed J, Beura LK, Bobr A, Astry B, Chicoine B, Kashem SW, Welty NE, Igyarto BZ, Wijeyesinghe S, Thompson EA, Matte C, Bartholin L, Kaplan A, Sheppard D, Bridges AG, Shlomchik WD, Masopust D, Kaplan DH (2016) Stromal cells control the epithelial residence of DCs and memory T cells by regulated activation of TGF-beta. Nat Immunol 17(4):414–421. https://doi.org/10.1038/ni.3396
Morrison LA, Zhu L, Thebeau LG (2001) Vaccine-induced serum immunoglobin contributes to protection from herpes simplex virus type 2 genital infection in the presence of immune T cells. J Virol 75(3):1195–1204. https://doi.org/10.1128/JVI.75.3.1195-1204.2001
Morrison RP, Feilzer K, Tumas DB (1995) Gene knockout mice establish a primary protective role for major histocompatibility complex class II-restricted responses in *Chlamydia trachomatis* genital tract infection. Infect Immun 63(12):4661–4668
Morrison SG, Morrison RP (2000) In situ analysis of the evolution of the primary immune response in murine *Chlamydia trachomatis* genital tract infection. Infect Immun 68(5):2870–2879. https://doi.org/10.1128/iai.68.5.2870-2879.2000
Morrison SG, Su H, Caldwell HD, Morrison RP (2000) Immunity to murine *Chlamydia trachomatis* genital tract reinfection involves B cells and CD4(+) T cells but not CD8(+) T cells. Infect Immun 68(12):6979–6987. https://doi.org/10.1128/iai.68.12.6979-6987.2000
Moyron-Quiroz JE, Rangel-Moreno J, Kusser K, Hartson L, Sprague F, Goodrich S, Woodland DL, Lund FE, Randall TD (2004) Role of inducible bronchus associated lymphoid

tissue (iBALT) in respiratory immunity. Nat Med 10(9):927–934. https://doi.org/10.1038/nm1091

Munoz-Suano A, Hamilton AB, Betz AG (2011) Gimme shelter: the immune system during pregnancy. Immunol Rev 241(1):20–38. https://doi.org/10.1111/j.1600-065X.2011.01002.x

Nardelli-Haefliger D, Wirthner D, Schiller JT, Lowy DR, Hildesheim A, Ponci F, De Grandi P (2003) Specific antibody levels at the cervix during the menstrual cycle of women vaccinated with human papillomavirus 16 virus-like particles. J Natl Cancer Inst 95(15):1128–1137. https://doi.org/10.1093/jnci/djg018

Oh JE, Iijima N, Song E, Lu P, Klein J, Jiang R, Kleinstein SH, Iwasaki A (2019) Migrant memory B cells secrete luminal antibody in the vagina. Nature. https://doi.org/10.1038/s41586-019-1285-1

Pan Y, Tian T, Park CO, Lofftus SY, Mei S, Liu X, Luo C, O'Malley JT, Gehad A, Teague JE, Divito SJ, Fuhlbrigge R, Puigserver P, Krueger JG, Hotamisligil GS, Clark RA, Kupper TS (2017) Survival of tissue-resident memory T cells requires exogenous lipid uptake and metabolism. Nature 543(7644):252–256. https://doi.org/10.1038/nature21379

Park CO, Fu X, Jiang X, Pan Y, Teague JE, Collins N, Tian T, O'Malley JT, Emerson RO, Kim JH, Jung Y, Watanabe R, Fuhlbrigge RC, Carbone FR, Gebhardt T, Clark RA, Lin CP, Kupper TS (2018a) Staged development of long-lived T-cell receptor αβ TH17 resident memory T-cell population to *Candida albicans* after skin infection. J Allergy Clin Immunol 142(2):647–662. https://doi.org/10.1016/j.jaci.2017.09.042

Park SL, Zaid A, Hor JL, Christo SN, Prier JE, Davies B, Alexandre YO, Gregory JL, Russell TA, Gebhardt T, Carbone FR, Tscharke DC, Heath WR, Mueller SN, Mackay LK (2018b) Local proliferation maintains a stable pool of tissue-resident memory T cells after antiviral recall responses. Nat Immunol 19(2):183–191. https://doi.org/10.1038/s41590-017-0027-5

Parr MB, Kepple L, McDermott MR, Drew MD, Bozzola JJ, Parr EL (1994) A mouse model for studies of mucosal immunity to vaginal infection by herpes simplex virus type 2. Lab Invest 70(3):369–380

Parr MB, Kepple L, Parr EL (1991) Langerhans cells phagocytose vaginal epithelial cells undergoing apoptosis during the murine estrous cycle. Biol Reprod 45(2):252–260. https://doi.org/10.1095/biolreprod45.2.252

Parr MB, Parr EL (1998) Mucosal immunity to herpes simplex virus type 2 infection in the mouse vagina is impaired by in vivo depletion of T lymphocytes. J Virol 72(4):2677–2685

Parr MB, Parr EL (1999) The role of gamma interferon in immune resistance to vaginal infection by herpes simplex virus type 2 in mice. Virology 258(2):282–294. https://doi.org/10.1006/viro.1999.9739

Permanyer M, Bosnjak B, Forster R (2018) Dendritic cells, T cells and lymphatics: dialogues in migration and beyond. Curr Opin Immunol 53:173–179. https://doi.org/10.1016/j.coi.2018.05.004

Perry LL, Feilzer K, Caldwell HD (1997) Immunity to *Chlamydia trachomatis* is mediated by T helper 1 cells through IFN-gamma-dependent and -independent pathways. J Immunol 158(7):3344–3352

Peters A, Pitcher LA, Sullivan JM, Mitsdoerffer M, Acton SE, Franz B, Wucherpfennig K, Turley S, Carroll MC, Sobel RA, Bettelli E, Kuchroo VK (2011) Th17 cells induce ectopic lymphoid follicles in central nervous system tissue inflammation. Immunity 35(6):986–996. https://doi.org/10.1016/j.immuni.2011.10.015

Phillips TA, Ni J, Hunt JS (2001) Death-inducing tumour necrosis factor (TNF) superfamily ligands and receptors are transcribed in human placentae, cytotrophoblasts, placental macrophages and placental cell lines. Placenta 22(8–9):663–672. https://doi.org/10.1053/plac.2001.0703

Plaks V, Birnberg T, Berkutzki T, Sela S, BenYashar A, Kalchenko V, Mor G, Keshet E, Dekel N, Neeman M, Jung S (2008) Uterine DCs are crucial for decidua formation during embryo implantation in mice. J Clin Invest 118(12):3954–3965. https://doi.org/10.1172/JCI36682

Poston TB, Darville T (2018) *Chlamydia trachomatis*: protective adaptive responses and prospects for a vaccine. Curr Top Microbiol Immunol 412:217–237. https://doi.org/10.1007/82_2016_6

Quinn TC, Overbaugh J (2005) HIV/AIDS in women: an expanding epidemic. Science 308 (5728):1582–1583. https://doi.org/10.1126/science.1112489

Randall TD, Carragher DM, Rangel-Moreno J (2008) Development of secondary lymphoid organs. Annu Rev Immunol 26:627–650. https://doi.org/10.1146/annurev.immunol.26.021607.090257

Reacher MH, Pe'er J, Rapoza PA, Whittum-Hudson JA, Taylor HR (1991) T cells and trachoma. Their role in cicatricial disease. Ophthalmology 98(3):334–341. https://doi.org/10.1016/s0161-6420(91)32290-5

Reinhardt RL, Bullard DC, Weaver CT, Jenkins MK (2003) Preferential accumulation of antigen-specific effector CD4 T cells at an antigen injection site involves CD62E-dependent migration but not local proliferation. J Exp Med 197(6):751–762. https://doi.org/10.1084/jem.20021690

Reinhardt RL, Khoruts A, Merica R, Zell T, Jenkins MK (2001) Visualizing the generation of memory CD4 T cells in the whole body. Nature 410(6824):101–105. https://doi.org/10.1038/35065111

Robertson SA, Care AS, Moldenhauer LM (2018) Regulatory T cells in embryo implantation and the immune response to pregnancy. J Clin Invest 128(10):4224–4235. https://doi.org/10.1172/JCI122182

Roth KL, Bhavanam S, Jiang H, Gillgrass A, Ho K, Ferreira VH, Kaushic C (2013) Delayed but effective induction of mucosal memory immune responses against genital HSV-2 in the absence of secondary lymphoid organs. Mucosal Immunol 6(1):56–68. https://doi.org/10.1038/mi.2012.48

Safaeian M, Falk RT, Rodriguez AC, Hildesheim A, Kemp T, Williams M, Morera L, Barrantes M, Herrero R, Porras C, Pinto L (2009a) Factors associated with fluctuations in IgA and IgG levels at the cervix during the menstrual cycle. J Infect Dis 199(3):455–463. https://doi.org/10.1086/596060

Safaeian M, Kemp T, Falk RT, Rodriguez AC, Hildesheim A, Williams M, Porras C, Herrero R, Pinto LA (2009b) Determinants and correlation of systemic and cervical concentrations of total IgA and IgG. Cancer Epidemiol Biomarkers Prev 18(10):2672–2676. https://doi.org/10.1158/1055-9965.EPI-09-0348

Sallusto F, Lenig D, Forster R, Lipp M, Lanzavecchia A (1999) Two subsets of memory T lymphocytes with distinct homing potentials and effector functions. Nature 401(6754):708–712. https://doi.org/10.1038/44385

Sautes-Fridman C, Petitprez F, Calderaro J, Fridman WH (2019) Tertiary lymphoid structures in the era of cancer immunotherapy. Nat Rev Cancer 19(6):307–325. https://doi.org/10.1038/s41568-019-0144-6

Schatz F, Guzeloglu-Kayisli O, Arlier S, Kayisli UA, Lockwood CJ (2016) The role of decidual cells in uterine hemostasis, menstruation, inflammation, adverse pregnancy outcomes and abnormal uterine bleeding. Hum Reprod Update 22(4):497–515. https://doi.org/10.1093/humupd/dmw004

Schenkel JM, Fraser KA, Beura LK, Pauken KE, Vezys V, Masopust D (2014a) T cell memory. Resident memory CD8 T cells trigger protective innate and adaptive immune responses. Science 346(6205):98–101. https://doi.org/10.1126/science.1254536

Schenkel JM, Fraser KA, Masopust D (2014b) Cutting edge: resident memory CD8 T cells occupy frontline niches in secondary lymphoid organs. J Immunol 192(7):2961–2964. https://doi.org/10.4049/jimmunol.1400003

Schenkel JM, Fraser KA, Vezys V, Masopust D (2013) Sensing and alarm function of resident memory CD8(+) T cells. Nat Immunol 14(5):509–513. https://doi.org/10.1038/ni.2568

Schenkel JM, Masopust D (2014) Tissue-resident memory T cells. Immunity 41(6):886–897. https://doi.org/10.1016/j.immuni.2014.12.007

Schiffer JT, Swan DA, Prlic M, Lund JM (2018) Herpes simplex virus-2 dynamics as a probe to measure the extremely rapid and spatially localized tissue-resident T-cell response. Immunol Rev 285(1):113–133. https://doi.org/10.1111/imr.12672

Schulz O, Hammerschmidt SI, Moschovakis GL, Forster R (2016) Chemokines and chemokine receptors in lymphoid tissue dynamics. Annu Rev Immunol 34:203–242. https://doi.org/10.1146/annurev-immunol-041015-055649

Seddon B, Tomlinson P, Zamoyska R (2003) Interleukin 7 and T cell receptor signals regulate homeostasis of CD4 memory cells. Nat Immunol 4(7):680–686. https://doi.org/10.1038/ni946

Sheridan BS, Pham QM, Lee YT, Cauley LS, Puddington L, Lefrancois L (2014) Oral infection drives a distinct population of intestinal resident memory CD8(+) T cells with enhanced protective function. Immunity 40(5):747–757. https://doi.org/10.1016/j.immuni.2014.03.007

Shimada S, Ebina Y, Iijima N, Deguchi M, Yamada H (2018) Decidual CD68(+) HLA-DR(+) CD163(-) M1 macrophages increase in miscarriages with normal fetal chromosome. Am J Reprod Immunol 79(2). https://doi.org/10.1111/aji.12791

Shin H, Iwasaki A (2012) A vaccine strategy that protects against genital herpes by establishing local memory T cells. Nature 491(7424):463–467. https://doi.org/10.1038/nature11522

Sigmundsdottir H, Butcher EC (2008) Environmental cues, dendritic cells and the programming of tissue-selective lymphocyte trafficking. Nat Immunol 9(9):981–987. https://doi.org/10.1038/ni.f.208

Skon CN, Lee JY, Anderson KG, Masopust D, Hogquist KA, Jameson SC (2013) Transcriptional downregulation of S1pr1 is required for the establishment of resident memory CD8 + T cells. Nat Immunol 14(12):1285–1293. https://doi.org/10.1038/ni.2745

Slifka MK, Amanna IJ (2019) Role of multivalency and antigenic threshold in generating protective antibody responses. Front Immunol 10:956. https://doi.org/10.3389/fimmu.2019.00956

Smolders J, Heutinck KM, Fransen NL, Remmerswaal EBM, Hombrink P, Ten Berge IJM, van Lier RAW, Huitinga I, Hamann J (2018) Tissue-resident memory T cells populate the human brain. Nat Commun 9(1):4593. https://doi.org/10.1038/s41467-018-07053-9

Soderberg KA, Linehan MM, Ruddle NH, Iwasaki A (2004) MAdCAM-1 expressing sacral lymph node in the lymphotoxin beta-deficient mouse provides a site for immune generation following vaginal herpes simplex virus-2 infection. J Immunol 173(3):1908–1913. https://doi.org/10.4049/jimmunol.173.3.1908

Soerens AG, Da Costa A, Lund JM (2016) Regulatory T cells are essential to promote proper CD4 T-cell priming upon mucosal infection. Mucosal Immunol 9(6):1395–1406. https://doi.org/10.1038/mi.2016.19

Song J, Lang F, Zhao N, Guo Y, Zhang H (2018) Vaginal lactobacilli induce differentiation of monocytic precursors toward langerhans-like cells: in vitro evidence. Front Immunol 9:2437. https://doi.org/10.3389/fimmu.2018.02437

Stanberry LR, Spruance SL, Cunningham AL, Bernstein DI, Mindel A, Sacks S, Tyring S, Aoki FY, Slaoui M, Denis M, Vandepapeliere P, Dubin G, GlaxoSmithKline Herpes Vaccine Efficacy Study Group (2002) Glycoprotein-D-adjuvant vaccine to prevent genital herpes. N Engl J Med 347(21):1652–1661. https://doi.org/10.1056/nejmoa011915

Stary G, Olive A, Radovic-Moreno AF, Gondek D, Alvarez D, Basto PA, Perro M, Vrbanac VD, Tager AM, Shi J, Yethon JA, Farokhzad OC, Langer R, Starnbach MN, von Andrian UH (2015) VACCINES. A mucosal vaccine against Chlamydia trachomatis generates two waves of protective memory T cells. Science 348(6241):aaa8205. https://doi.org/10.1126/science.aaa8205

Steinbach K, Vincenti I, Kreutzfeldt M, Page N, Muschaweckh A, Wagner I, Drexler I, Pinschewer D, Korn T, Merkler D (2016) Brain-resident memory T cells represent an autonomous cytotoxic barrier to viral infection. J Exp Med 213(8):1571–1587. https://doi.org/10.1084/jem.20151916

Steinfelder S, Rausch S, Michael D, Kuhl AA, Hartmann S (2017) Intestinal helminth infection induces highly functional resident memory CD4(+) T cells in mice. Eur J Immunol 47(2):353–363. https://doi.org/10.1002/eji.201646575

Strutt TM, Dhume K, Finn CM, Hwang JH, Castonguay C, Swain SL, McKinstry KK (2018) IL-15 supports the generation of protective lung-resident memory CD4 T cells. Mucosal Immunol 11(3):668–680. https://doi.org/10.1038/mi.2017.101

Takamura S, Yagi H, Hakata Y, Motozono C, McMaster SR, Masumoto T, Fujisawa M, Chikaishi T, Komeda J, Itoh J, Umemura M, Kyusai A, Tomura M, Nakayama T, Woodland DL, Kohlmeier JE, Miyazawa M (2016) Specific niches for lung-resident memory CD8 + T cells at the site of tissue regeneration enable CD69-independent maintenance. J Exp Med 213(13):3057–3073. https://doi.org/10.1084/jem.20160938

Teijaro JR, Turner D, Pham Q, Wherry EJ, Lefrancois L, Farber DL (2011) Cutting edge: tissue-retentive lung memory CD4 T cells mediate optimal protection to respiratory virus infection. J Immunol 187(11):5510–5514. https://doi.org/10.4049/jimmunol.1102243

Turner DL, Goldklang M, Cvetkovski F, Paik D, Trischler J, Barahona J, Cao M, Dave R, Tanna N, D'Armiento JM, Farber DL (2018) Biased generation and in situ activation of Lung tissue-resident memory CD4 T cells in the pathogenesis of allergic asthma. J Immunol 200 (5):1561–1569. https://doi.org/10.4049/jimmunol.1700257

Tyssen D, Wang YY, Hayward JA, Agius PA, DeLong K, Aldunate M, Ravel J, Moench TR, Cone RA, Tachedjian G (2018) Anti-HIV-1 activity of lactic acid in human cervicovaginal fluid. mSphere 3(4). https://doi.org/10.1128/msphere.00055-18

Ugur M, Schulz O, Menon MB, Krueger A, Pabst O (2014) Resident CD4 + T cells accumulate in lymphoid organs after prolonged antigen exposure. Nat Commun 5:4821. https://doi.org/10.1038/ncomms5821

van de Pavert SA, Mebius RE (2010) New insights into the development of lymphoid tissues. Nat Rev Immunol 10(9):664–674. https://doi.org/10.1038/nri2832

Vince G, Shorter S, Starkey P, Humphreys J, Clover L, Wilkins T, Sargent I, Redman C (1992) Localization of tumour necrosis factor production in cells at the materno/fetal interface in human pregnancy. Clin Exp Immunol 88(1):174–180. https://doi.org/10.1111/j.1365-2249.1992.tb03059.x

Wakim LM, Gupta N, Mintern JD, Villadangos JA (2013) Enhanced survival of lung tissue-resident memory CD8(+) T cells during infection with influenza virus due to selective expression of IFITM3. Nat Immunol 14(3):238–245. https://doi.org/10.1038/ni.2525

Wakim LM, Woodward-Davis A, Bevan MJ (2010) Memory T cells persisting within the brain after local infection show functional adaptations to their tissue of residence. Proc Natl Acad Sci U S A 107(42):17872–17879. https://doi.org/10.1073/pnas.1010201107

Walsh DA, Borges da Silva H, Beura LK, Peng C, Hamilton SE, Masopust D, Jameson SC (2019) The functional requirement for CD69 in establishment of resident memory CD8(+) T cells varies with tissue location. J Immunol 203(4):946–955. https://doi.org/10.4049/jimmunol.1900052

Wang S, Fan Y, Brunham RC, Yang X (1999) IFN-gamma knockout mice show Th2-associated delayed-type hypersensitivity and the inflammatory cells fail to localize and control chlamydial infection. Eur J Immunol 29(11):3782–3792. https://doi.org/10.1002/(SICI)1521-4141(199911)29:11%3c3782:AID-IMMU3782%3e3.0.CO;2-B

Watanabe R, Gehad A, Yang C, Scott LL, Teague JE, Schlapbach C, Elco CP, Huang V, Matos TR, Kupper TS, Clark RA (2015) Human skin is protected by four functionally and phenotypically discrete populations of resident and recirculating memory T cells. Sci Transl Med 7(279):279ra239. https://doi.org/10.1126/scitranslmed.3010302

Wilk MM, Misiak A, McManus RM, Allen AC, Lynch MA, Mills KHG (2017) Lung CD4 tissue-resident memory T cells mediate adaptive immunity induced by previous infection of mice with *Bordetella pertussis*. J Immunol 199(1):233–243. https://doi.org/10.4049/jimmunol.1602051

Wira CR, Fahey JV, Rodriguez-Garcia M, Shen Z, Patel MV (2014) Regulation of mucosal immunity in the female reproductive tract: the role of sex hormones in immune protection against sexually transmitted pathogens. Am J Reprod Immunol 72(2):236–258. https://doi.org/10.1111/aji.12252

Wira CR, Rodriguez-Garcia M, Patel MV (2015) The role of sex hormones in immune protection of the female reproductive tract. Nat Rev Immunol 15(4):217–230. https://doi.org/10.1038/nri3819

Yeaman GR, Collins JE, Fanger MW, Wira CR, Lydyard PM (2001) CD8 + T cells in human uterine endometrial lymphoid aggregates: evidence for accumulation of cells by trafficking. Immunology 102(4):434–440. https://doi.org/10.1046/j.1365-2567.2001.01199.x

Yeaman GR, Guyre PM, Fanger MW, Collins JE, White HD, Rathbun W, Orndorff KA, Gonzalez J, Stern JE, Wira CR (1997) Unique CD8 + T cell-rich lymphoid aggregates in human uterine endometrium. J Leukoc Biol 61(4):427–435

Zhang N, Bevan MJ (2013) Transforming growth factor-beta signaling controls the formation and maintenance of gut-resident memory T cells by regulating migration and retention. Immunity 39(4):687–696. https://doi.org/10.1016/j.immuni.2013.08.019

Zhang Y, Liu Z, Tian M, Hu X, Wang L, Ji J, Liao A (2018) The altered PD-1/PD-L1 pathway delivers the 'one-two punch' effects to promote the Treg/Th17 imbalance in pre-eclampsia. Cell Mol Immunol 15(7):710–723. https://doi.org/10.1038/cmi.2017.70

Zhao X, Deak E, Soderberg K, Linehan M, Spezzano D, Zhu J, Knipe DM, Iwasaki A (2003) Vaginal submucosal dendritic cells, but not langerhans cells, induce protective Th1 responses to herpes simplex virus-2. J Exp Med 197(2):153–162. https://doi.org/10.1084/jem.20021109

Zhu J, Hladik F, Woodward A, Klock A, Peng T, Johnston C, Remington M, Magaret A, Koelle DM, Wald A, Corey L (2009) Persistence of HIV-1 receptor-positive cells after HSV-2 reactivation is a potential mechanism for increased HIV-1 acquisition. Nat Med 15(8):886–892. https://doi.org/10.1038/nm.2006

Zhu J, Peng T, Johnston C, Phasouk K, Kask AS, Klock A, Jin L, Diem K, Koelle DM, Wald A, Robins H, Corey L (2013) Immune surveillance by CD8alphaalpha + skin-resident T cells in human herpes virus infection. Nature 497(7450):494–497. https://doi.org/10.1038/nature12110

Tertiary Lymphoid Organs in Rheumatoid Arthritis

Felice Rivellese, Elena Pontarini, and Costantino Pitzalis

Contents

Abstract Rheumatoid Arthritis (RA) is a chronic systemic autoimmune disease. RA mainly affects the joints, with inflammation of the synovial membrane, characterized by hyperplasia, neo-angiogenesis, and immune cell infiltration that drives local inflammation and, if untreated, can lead to joint destruction and disability. In parallel to the well-known clinical heterogeneity, the underlying synovitis can also be significantly heterogeneous. In particular, in about 40% of patients with RA, synovitis is characterized by a dense lymphocytic infiltrate that can acquire the features of fully functional tertiary lymphoid organs (TLO). These structures amplify autoimmunity and inflammation locally associated with worse prognosis and potential implications for treatment response. Here, we will review the current knowledge on TLO in RA, with a focus on their pathogenetic and clinical relevance.

F. Rivellese · E. Pontarini · C. Pitzalis (✉)
Barts and the London School of Medicine & Dentistry, Centre for Experimental Medicine & Rheumatology, William Harvey Research Institute, John Vane Science Centre, London, UK
e-mail: c.pitzalis@qmul.ac.uk

Current Topics in Microbiology and Immunology (2020) 426: 119–142
https://doi.org/10.1007/82_2020_216

Published Online: 02 June 2020

1 Introduction

Rheumatoid Arthritis (RA) is the most common autoimmune disease, affecting up to 1% of the population worldwide (Smolen et al. 2016). Although RA is well recognized as a systemic disease, its main feature is the chronic inflammation of the synovial membrane, which is characterized by infiltration of immune cells, cellular hyperplasia, and neo-angiogenesis (McInnes and Schett 2011, 2017). Ongoing synovitis and its corresponding clinical features of joint pain and swelling are the main causes of functional disability in patients with RA. Despite the availability of effective medications, in a large proportion of patients, the treatments fail to control the inflammatory response. When un-optimally controlled, synovial inflammation can progress and ultimately lead to joint destruction and permanent disability. Such inconsistent response to treatment has been attributed at least in part to the clinical and physiopathological heterogeneity of RA. In fact, similar to other autoimmune diseases, under the umbrella of RA, we are grouping a diverse spectrum of patients with different clinical features, which are mirrored by significant differences in terms of pathogenesis and, therefore, variable response to targeted treatments. For example, it is well recognized that the positivity for anti-citrullinated protein antibodies (ACPA) identifies a group of patients—around 70%- with a clinical phenotype of highly aggressive and destructive disease (Willemze et al. 2012). In line with its marked clinical heterogeneity, a variable degree of immune cell infiltration has been described in the synovia of RA patients and has been recently linked to distinct clinical features, including disease severity, progression, and treatment response.

2 The Synovial Membrane as Site of Inflammation in RA

The main physiopathological feature of RA is the inflammation of the synovial membrane (SM). In physiological condition, the SM is a composed by an intimal layer formed of synoviocytes, also known as fibroblast-like synoviocytes (FLS), which are specialized fibroblast-like cells with the main function of producing the synovial fluid that lubricates and nourish the avascular articular surfaces. Below the thin layer of FLS, there is a sub-intimal layer composed by connective tissues, scattered infiltrating macrophage-like cells, and blood vessels. During RA, the synovial membrane undergoes the following changes: (i) infiltration of immune cells, including cells of innate (e.g., macrophages, natural killer [NK] cells, innate lymphoid cells, dendritic cells, mast cells) and adaptive immunity (e.g., B and T lymphocytes, plasma cells); (ii) proliferation of FLS, leading to the thickening of the intimal layer, and (iii) growth of new blood vessels (neo-angiogenesis) which further sustains the infiltration of immune cells, thus facilitating the perpetuation of the inflammatory response. Despite the enormous advancements in our understanding of the pathogenesis of RA, leading us to recognize a number of genetic and environmental factors contributing to its pathogenesis, the initial trigger of synovial inflammation is currently unknown. Also, we do not know whether the first hit happens directly in the joints or somewhere else, such as the lungs or other organs. However, once the inflammatory response is triggered and gets perpetuated, synovitis represents the main feature of RA, thus the study of synovial inflammation is of utmost importance to improve our understanding of RA (Pitzalis et al. 2013).

2.1 Histological Patterns of Synovial Inflammation

The infiltration of immune cells is one of the main features of RA synovitis. In line with the clinical heterogeneity of the disease, a variable degree of immune cell infiltration in synovia has been described. Despite the complexity and partial overlap of immune cell infiltration, the parallel study of large numbers of synovial samples from patients with early untreated RA (Humby et al. 2019) has allowed to describe three distinct groups based on the patterns of immune cell infiltration in synovia: (1) lympho-myeloid, dominated by lymphoid lineage infiltration (T cells, B cells, plasma cells) in addition to myeloid cells; (2) diffuse-myeloid, with myeloid lineage predominance but poor in B cells/plasma cells and (3) pauci-immune, characterized by scanty immune cells and prevalent stromal cells. Within the lympho-myeloid group, the infiltrating B cells, T cells, and plasma cells often organize into aggregates that resemble the lymphoid follicles of secondary lymphoid organs, acquiring features such as segregation of T cells and B cells, the presence of high endothelial venules (HEVs), and follicular dendritic cells (FDCs) networks. Although TLO can also be detected at extra-articular sites, including the lungs (Barone et al. 2015) and bone marrow (Bugatti et al. 2005) of RA patients, they mainly form within the sublining of the synovial tissue, where they have been described in about 40% of patients with early untreated RA (Pitzalis et al. 2013). A representative example of TLO is offered in Fig. 1a–c, including a schematic representation of their organization in Fig. 1d, with additional details in Fig. 2. In the next paragraphs, we will describe the ontogeny of tertiary lymphoid organs in RA, their functions, and their correlation with clinical features and disease prognosis, including response to treatment.

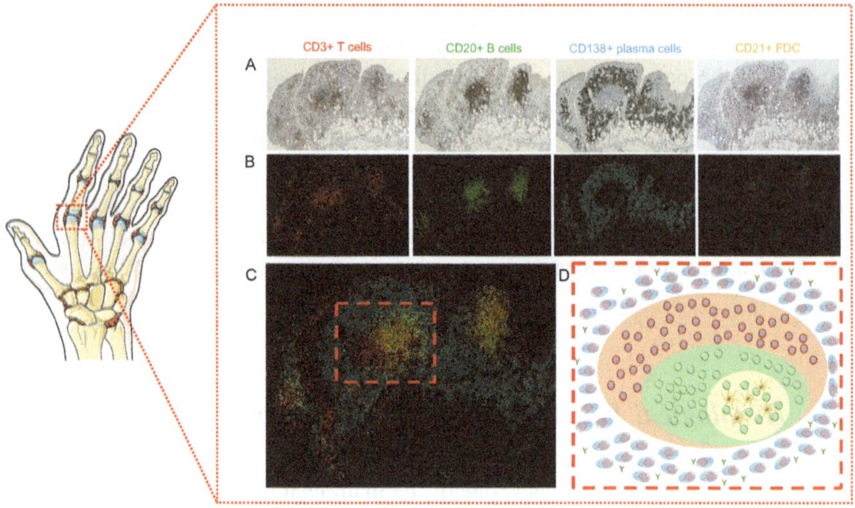

Fig. 1 Tertiary lymphoid organs in synovia. **a** Immunohistochemical staining of synovial membrane, **b** color deconvolution of the images in (**a**), **c** overlap of the above images, and **d** schematic representation of the organization of TLO in synovia, with FDC in yellow, B cells in green, T cells in red, and plasma cells in blue

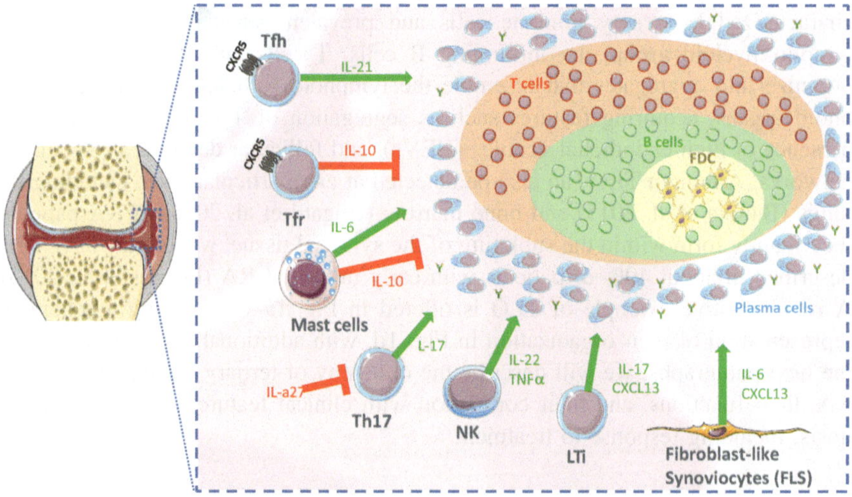

Fig. 2 Schematic representation of synovial TLOs and immune cells contributing to their development. Tfh = T follicular helper cells; Tfr = T follicular regulatory cells, Th17 = T helper cells 17, NK = Natural killer cells; LTi = Lymphoid tissue inducer

3 Synovial Tertiary Lymphoid Organs in RA

3.1 The Development and Regulation of Synovial Tertiary Lymphoid Organs in RA

3.1.1 Chemokine and Lymphotoxin Beta

One of the initial steps in the formation of TLO is the infiltration of lymphoid cells into the synovia, which is driven by the inflammatory milieu produced by FLS and innate immunity cells. As the inflammatory process becomes chronic, however, a number of specific mediators are required for the formation of TLO, such as lymphotoxin-β (LTβ), CXCL13, CCL19, and CCL21 (Corsiero et al. 2012). The development of TLO largely mirrors the ontogeny of secondary lymphoid organs, thus most of our knowledge on TLO development is derived from the study of secondary lymphoid organs, where animal models have identified a number of stimuli which are essential for the development of secondary lymphoid organs (Randall et al. 2008; Drayton et al. 2006). Among these stimuli, the *primum movens* has been recognized to be the production of lymphotoxin-β from so-called lymphoid tissue inducer (LTi) cells (Bar-Ephraim and Mebius 2016), which in turns leads to the production of lymphoid chemokines (CXCL13, CCL19, and CCL21) from lymphoid tissue organizers and mesenchymal cells. Although the presence of many of these lymphogenic stimuli has been confirmed in TLO in rheumatoid synovium (Bugatti et al. 2014; Manzo et al. 2007), the initial trigger of TLO formation in RA has not been identified. Several immune cells have been shown to be a source of lymphoid chemokines, and some of these key cells are represented in Fig. 2b. Among the various mediators, CXCL13 produced by follicular dendritic cells (Takemura et al. 2001) and other immune cells plays a pivotal role in determining the spatial organization of TLO, inducing the segregation of B cells within the germinal center, which is an essential drive for affinity maturation and antigenic selection (De Silva and Klein 2015). In line with its pivotal role, serum levels of CXCL13 have been associated with the presence of synovial TLO in patients with RA (Bugatti et al. 2014; Dennis et al. 2014).

3.1.2 T Follicular Helper Cells

In recent years, a specialized class of T helper cells, named T follicular helper cells (Tfh), has been recognized for their central role in sustaining B cell activation and differentiation in the germinal center (GC) reactions in secondary lymphoid organs. Tfh cells are specialized T helper cells that upon priming by antigen presenting cells (APCs) acquire the expression of CXCR5, the receptor for CXCL13, enabling them to migrate into the B cell area of GC.

The ectopic expression of CXCL13 has been described in RA synovium (Manzo et al. 2005, 2008) and has been shown to induce TLO formation and recruits B cells to non-lymphoid tissues in mice (Luther et al. 2000). In fully formed GC, Tfh cells support somatic hypermutation of auto-reactive B cells and plasmablast generation directly in the diseased tissues mainly through the production of IL-21. The latter is Tfh signature cytokine, known to be a potent cofactor for B cell survival, proliferation and plasma cell differentiation, in particular in the context of CD40 co-stimulation and in synergy with B cell activating factor (Karnell and Ettinger 2012; Liu et al. 2015).

Importantly, because of the role of IL-21 and Tfh cells in supporting GC response, they have been implicated in the development of TLO in rheumatic autoimmune diseases, including RA, as represented in Fig. 2.

Data in animal models of arthritis identified a number of Tfh-associated markers during the development of inflammatory arthritis. In particular, CXCR5 has been shown to be an essential factor for the development of inflammatory arthritis: CXCR5-deficient animals or lacking CXCR5 on T cells are resistant to RA, showing impaired GC response (Moschovakis et al. 2017). Also, selective deficiency in T helper cells of SLAM-associated protein (SAP), required for the B/T cell interaction, thus essential for Tfh differentiation, protects mice from RA, further supporting the pathogenic role of ectopic GC formation (McCausland et al. 2014).

In parallel, IL-21 and its receptor are highly expressed in synovial tissue of patients with RA (Jüngel et al. 2004; Kwok et al. 2012), and increased IL-21 expression is associated with synovial TLO (Jones et al. 2015). IL-21R up-regulation is mainly described on macrophages and fibroblast with an activated phenotype (Jüngel et al. 2004), and IL-21 has been involved in the development of articular damage by promoting both osteoclastogenesis (Kwok et al. 2012) and metalloproteinase release by fibroblast-like synoviocytes (Xing et al. 2016). Finally, Tfh cells are also enriched in the synovia of patients with RA, while almost absent in osteoarthritis and normal synovium (Penatti et al. 2017; Chu et al. 2014).

In addition to conventional CXCR5+ Tfh cells, a population of T helper cells lacking CXCR5 expression and producing CXCL13 has been also described in the synovia of RA patients (Manzo et al. 2008). A recent breakthrough publication has shed new light on these cells, which have been re-named as PD1+ CXCR5—T peripheral helper cells (Tph), since they have been found in the synovia but also in the peripheral blood of patients with RA and their ability to induce the activation of B cells has been confirmed in vitro (Rao et al. 2017). Similar to GC-Tfh, these cells are an important source of CXCL13, support synovial B cell proliferation and activation through IL-21 production and SLAMF5 receptor ligation, and co-localize with B cells in synovial TLO (Rao et al. 2017). Although Tfh and Tph cells share the main markers, the tissue localization, and the ability to support B cell activation, it is unclear whether Tph in RA are Tfh cells with impaired CXCR5 expression, or a more distantly related cell type. Despite the evidence of Tfh and Tph contribution to the pathogenesis of RA, the functional link between these cells and TLO formation remains to be elucidated, as well as their contribution to the local production of autoantibodies within TLO.

Additionally, although the enrichment of Tfh in RA synovium has been well described, there are conflicting data regarding circulating Tfh cell frequency [comprehensively reviewed in (Gensous et al. 2018)]. Some authors reported IL-21 directly correlating with the frequency of Tfh-like cells, with IL-21 level and number of Tfh-like cells associated with higher titer of anti-CCP antibodies and disease activity score in RA (Ma et al. 2012). The circulating counterpart shares phenotypic and functional features with tissue Tfh cells, except for the expression of prototypical Tfh transcription factor B cell lymphoma protein 6 (Bcl-6), but their biology is still poorly defined. Data from SAP-deficient mice show how these cells are committed to Tfh lineage and are generated prior the GC response (He et al. 2013; Tsai and Yu 2014). Moreover, it is still unclear if circulating Tfh can reflect an ongoing humoral activity.

3.1.3 Other Pro-inflammatory Cytokines and Cells

It is now clear that a number of other pro-inflammatory cytokines, such as IL-17, IL-21, IL-22, IL-23, and TNFα, are also critical for lymphoid neogenesis in autoimmune diseases (Jones and Jones 2016).

The IL-23–IL-17 pathway has been involved in the initiation and perpetuation of TLO, and several cells of the innate and adaptive immunity are able to produce IL-17. In particular, a subset of adult innate lymphoid cells [type-3 innate lymphoid cells (ILC3 cells)] can produce IL-17 in the initial phases of TLO formation (Sawa et al. 2010). Accordingly, IL-17 positive cells are observed in the proximity of TLO in RA synovia (Chabaud et al. 1999), and the activation of the IL-23–IL-17 pathway correlates with the presence of synovial TLO (Cañete et al. 2011).

Another important aspect is the potential plasticity between other T helper subsets and the Tfh. In fact, several other subsets, including Th17 cells, Th1 and Th2, have been described to acquire Tfh-like phenotype (Ueno et al. 2015). For example, Tfh2 and Tfh17, but not Tfh1, are able to secrete IL-21 and induce naïve B cells to secrete class-switched immunoglobulin (Ig) (Morita et al. 2011).

Within RA synovium, proliferation of fibroblast-like synoviocytes is sustained by IL-22, a cytokine required for the development and maintenance of TLO. IL-22 role in ectopic lymphoneogenesis comes from data in experimental models of inducible TLO in salivary glands, mimicking TLO formation in Sjogren's syndrome salivary glands. In this animal model, IL-22 is able to directly induce CXCL13 production in a subset of GP38+ stromal cells through phosphorylation of signal transduced and activator of transcription 3 (STAT3) (Barone et al. 2015). Once lymphocytes are recruited, IL-22, together with LTα1β2, supports also proliferation of a population of podoplanin (pdpn)-positive stromal cells, over-expressing IL21R, into a network of immunofibroblasts that are able to support the earliest phases of TLS establishment (Nayar et al. 2019) in the same model. In RA synovium, IL-22 expression and IL-22 receptor on fibroblast-like synoviocytes have been reported (Ikeuchi et al. 2005), suggesting its contribution to the maintenance of TLO. In particular, IL-22 expression is increased in cells expressing the long isoform of complement receptor type 2 (Cr2, also known as CD21) (Cañete et al. 2011), usually present in networks of stromal-derived follicular dendritic cells (FDCs), that contribute to the presentation of immune complexes necessary to generate activated B cells, in TLO. In synovial tissue, IL-22 is also produced by NK cells (Zhu et al. 2015). NK cells are innate immune lymphocytes with cytolytic and immune-regulatory activities representing a significant proportion (8–25%) of immune infiltration in synovial fluid of RA patients, identified in the joints in the early stage of RA development (Tak et al. 1994). Initially, NK cells were described in RA pathogenesis for their production of cytotoxic serin protease granzyme-A and B and pro-inflammatory cytokines, such as IL-1 and TNFα as dominant mediators of proliferative synovitis in RA (Klimiuk et al. 1997), supporting ocleoclastogenesis and thus involved in the development of articular damage (Kotake et al. 2001). In fact, increased production of IFNγ and TNFα characterizes synovial fluid NK cells of erosive RA patients with joint damage in comparison with non-erosive RA (Yamin et al. 2019). Recent evidence suggests that NK cells may support TLO maintenance within RA synovium as a subset of NK cells expressing a natural cytotoxicity receptor NKp44 which is able to produce IL-22 (Zhu et al. 2015). NKp44+ NK cells are enriched in both peripheral blood and synovium of RA patients secreting IL-22 and TNFα, which in vitro studies showed to support RA FLS proliferation (Ren et al. 2011), through the activation of STAT3 pathway (Zhu et al. 2015). IL-22 induced proliferation of synovial fibroblast, an effect that was inhibited by neutralizing antibodies targeting IL-22 and TNFα (Ren et al. 2011). Thus, NK cells may participate in TLO organization supporting the proliferation of synovial fibroblasts responsible for the local secretion of chemoattractant molecules and, as consequence, lymphocytes recruitment.

In addition to cells of the adaptive immunity, many other innate immunity cells and the stromal compartment have been shown to contribute to the development of synovial TLO (Barone et al. 2016).

Fibroblast-like synoviocytes (FLS), for example, have been shown to produce the T cell/dendritic cell chemoattractant CCL21 (Manzo et al. 2007) and express CXCL12 and IL-7, involved in immune cell retention and lymphoid-like microanatomical organization (Timmer et al. 2007; Bradfield et al. 2003).

Recently, we have also shown a strong association between synovial mast cells (MCs) and the presence of TLO in a large cohort of patients with early RA (Rivellese et al. 2018). MCs were also found to induce B cell activation and differentiation in vitro, including the production of ACPA autoantibodies. Finally, in animal models of inducible TLO (IL27R knockout), we confirmed the association of MCs with TLO. Overall, this points out to the relevance of MCs as potential contributors to the formation of TLO, although additional studies are needed to confirm their functional relevance (Rivellese et al. 2017, 2019b).

3.1.4 Negative Regulators of TLO Including Tfr

In addition to the mediators and pathways acting as positive regulators of TLO, several cells and cytokines have been characterized as negative regulators of TLO development.

For example, IL-27, an heterodimeric cytokines part of the IL-12 family (Yoshida and Hunter 2015), has been recently identified as a negative regulator of TLO. In fact, animals with knockout of the IL27Rα develop a severe form of antigen-induced arthritis, including the development of synovial TLO (Jones et al. 2015). Importantly, synovial TLO are not normally produced in animal models of arthritis; thus, the identification of these structures in IL-27Rα knockout animals points to the relevance of IL-27 as a regulator of TLO development. Accordingly, in patients with RA, IL-27 was found to be inversely correlated with TLO and with TLO-related gene signatures. Finally, both in clinical and experimental arthritis, synovial TLO coincided with an increased local expression of cytokines and transcription factors of the Th17 and T follicular helper (Tfh) cell lineages, where IL-27 is able to inhibit the differentiation of Th17 cells, in line with previous evidence (Stumhofer et al. 2006).

As local counterpart of the circulating T regulatory cells, T follicular regulatory cells (Tfr) have been recently described within GCs, including GCs in TLO. Tfr cells are able to prevent the differentiation of auto-reactive B cells (Wu et al. 2016; Botta et al. 2017), by regulating Tfh cells, but also by directly inhibiting B cell activation (Wing et al. 2014).

Although the relevance of Tfr cells in the regulation of GCs in animal models is well established (Linterman et al. 2011), the involvement of Tfr cells in human autoimmune disease, including RA, is still unclear.

Several studies have reported decreased levels of Tfr in patients with active RA and, accordingly, negative correlations with autoantibodies and disease activity (Romão et al. 2018; Niu et al. 2018). On the other hand, increased levels of Tfr were found in patients who were in remission (Liu et al. 2018). Using animal models of autoimmunity with spontaneous development of GCs, IL-21 was shown to induce an unbalance between Tfh and Tfr, increasing the formation of GCs, while administration of Tfr was able to restore Tfh:Tfr ration and suppress GC responses (Ding et al. 2014).

Another group found that the resolution of collagen-induced arthritis following administration of intravenous immunoglobulins was accompanied by an increase of Tfr cells (Lee et al. 2014). Taken together, this suggests that the reduction of circulating Tfr cells is associated to the development of RA and that restoration of Tfr cells could potentially improve autoimmune responses.

In line with this, monitoring the ratio between Tfr and Tfr could be useful in patients with RA, as confirmed by a several observations (Niu et al. 2018; Wang et al. 2019).

As for the function of Tfr in RA, these cells have been shown to have suppressive effects in vitro, which were enhanced in patients in remission (Liu et al. 2018). However, it has also been speculated that Tfr in autoimmune diseases might be functionally deficient (Fonseca et al. 2017).

3.2 *The Function of Tertiary Lymphoid Organs in RA*

As TLO mirrors secondary lymphoid organs in their ontogeny and maturation, it is expected that they also recapitulate the main functions of secondary lymphoid organs, which is supporting germinal centers (GC) reactions toward maturation of B cells and antibody production.

Within a considerable proportion of TLO forming in rheumatoid synovium, ectopic GC reactions take place similar to secondary lymphoid organs (Bombardieri et al. 2017). Many of RA-associated autoantibodies are high affinity IgG (e.g., ACPA) (van Delft and Huizinga 2020), and B cells forming TLO are auto-reactive and somatically mutated (Humby et al. 2009), indicating the involvement of a GC response in RA progression. Indeed, TLO in RA synovium can display functional features of germinal centers, like the expression of the enzyme activation-induced cytidine deaminase (AID) involved in in situ B cell affinity maturation and clonal selection (Humby et al. 2009).

Accordingly, the analysis of B cells isolated from the synovia of patients with RA has confirmed the generation of synovial plasma cells from locally activated B cells (Scheel et al. 2011), and the local production of class-switched autoantibodies in rheumatoid synovium has been demonstrated (Humby et al. 2009). Also, we have recently demonstrated that the presence of synovial TLO in early untreated RA is associated with autoantibody positivity (Humby et al. 2019). Interestingly, this is in contrast with previous data that failed to show an association between TLO and autoantibody positivity (Thurlings et al. 2008). Recently, comparing two large cohorts of patients with early and established RA, we were able to confirm the strong association between TLO and autoantibody positivity in early RA that could not be observed in established RA, thus explaining the previous findings, possibly because of treatment effect or other biases from long-standing diseases (Rivellese et al. 2019a).

Importantly, the initiation of a germinal center reaction requires antigen presentation to B cells. In RA, the aberrant immune response against citrullinated proteins culminating in the production of anti-citrullinated protein antibodies (ACPA) is well recognized as a key pathogenetic feature (Derksen et al. 2017).

Accordingly, citrullinated proteins have been described in the synovia of RA patients (Baeten et al. 2001) together with PAD enzymes, which are responsible for citrullination (De Rycke et al. 2005). The specificity of synovial citrullinated protein has been challenged (Vossenaar et al. 2004), but this does not come as a surprise since citrullination and other post-translational modifications of proteins are recognized as physiological processes (Trouw et al. 2017). On the contrary, the aberrant immune response to modified proteins represents the hallmark of RA, and accordingly, the local production of ACPA in synovia has been confirmed (Humby et al. 2009; Amara et al. 2013; Masson-Bessière et al. 2000). Finally, several groups have been able to isolate ACPA-producing B cell clones from the synovia and synovial fluid of patients with RA (Germar et al. 2019; Corsiero et al. 2016, 2018).

3.3 The Clinical Relevance of Tertiary Lymphoid Organs in RA

3.3.1 TLO and Disease Severity

Early studies on the analysis of synovial membrane relied on the use arthroscopy to obtain synovial samples. These analyses pointed out a marked heterogeneity in terms of synovial inflammation, particularly in the degree of immune cell infiltration, with the description of aggregates of lymphoid cells in a proportion of patients. However, when looking for an association with clinical features, these studies yielded contradictory results: some found an association of lymphoid aggregates with disease severity and autoantibody positivity (Humby et al. 2019; Bugatti et al. 2014; Orr et al. 2017) and others did not (Thurlings et al. 2008; Cantaert et al. 2008; Van De Sande et al. 2011) (Table 1). These inconsistencies could be explained by a number of biases: (i) the exclusive analysis of large joints, in which there can be commonly overlapping osteoarthritis and are not the most representative of the inflammatory process in RA (Linn-Rasker et al. 2007) (ii) the inclusion of patients with long-standing disease, with the obvious bias of treatment and disease duration, and (iii) the lack of a gold standard for the histological assessment of immune cell infiltration (Humby et al. 2016).

The development of minimally invasive techniques such as ultrasound-guided synovial biopsies has overcome most of these limitations, as it made possible to obtain synovial tissues from small joints of a large cohort of patients with early RA and, very importantly, prior to treatment start. Thus, it is not surprising that the recently published analyses on this cohort highlighted a strong association with disease severity and autoantibody positivity (Humby et al. 2019). Interestingly, a direct comparison of early and established RA, using a validated semi-quantitative score for the assessment of B cells, showed that while in early RA the presence of B cell-rich synovitis was associated with disease severity, this was not the case in established RA, possibly explaining the discrepancies from previous studies analyzing patient with different disease duration (Rivellese et al. 2019a).

When analyzing patients with early untreated RA, our group has recently shown that patients with a synovial lympho-myeloid pathotype, characterized by the presence of B and T cell aggregates, have significantly higher disease severity, autoantibody positivity, and baseline erosive load (Humby et al. 2019). Furthermore, molecular analyses showed that myeloid- and lymphoid-associated genes strongly correlate with disease activity and acute phase reactants. Another

Table 1 Association of TLO with disease severity and clinical phenotype in RA

Author and year	References	Population	Joints biopsied and procedure	Treatment (if any)	Time points	Analyses	Results
Van Oosterhout 2008	Van Oosterhout et al. (2008)	57 RA	Knee arthroscopy	N.a.	Biopsy at time 0	IHC	ACPA + patients showed higher mean number of infiltrating lymphocytes and higher rate of local joint destruction
Van de Sande 2011	Van De Sande et al. (2011)	93 (24 RA)	Knee arthroscopy	sDMARDs	Biopsy at 0 (93) and 6 months (17)	IHC	Lymphoid neogenesis present in 36% of all patients, associated with the degree of synovial inflammation, but not specific of RA. No relationship between the presence of lymphocyte aggregates at baseline and definitive diagnosis or clinical outcome after follow-up
De Hair 2013	De Hair et al. (2013)	55 seropositive individuals without clinical evidence of arthritis	Knee arthroscopy	N.a.	Biopsy at time 0	IHC	CD3 T cell numbers in the biopsy tissue showed a borderline association with subsequent development of clinically manifest arthritis. CD8 T cells were associated with ACPA positivity
Gómez-Puerta 2013	Gómez-Puerta et al. (2013)	83 RA	Knee arthroscopy	N.a.	Biopsy at time 0	IHC	No significant differences in clinical variables, acute phase reactants, synovial cell infiltrate or lymphoid neogenesis (LN) between ACPA positive and negative patients
Orr 2017	Orr et al. (2017)	123 RA	Knee arthroscopy	sDMARDs and bDMARDs	Biopsy at time 0	IHC	ACPA + RA patients were characterized by significantly higher levels of CD19+ B cells and CD3+ and CD8+ T cells. Levels of lymphoid aggregates of CD19+ B cells and serum CXCL13 levels were significantly

(continued)

Table 1 (continued)

Author and year	References	Population	Joints biopsied and procedure	Treatment (if any)	Time points	Analyses	Results
							higher in ACPA + patients. EULAR response was significantly associated with the level of CD3+ T cell infiltrates, while CD68+ macrophage and CD8+ T cell levels were predictive of the response to tumor necrosis factor inhibitors
Humby 2019	Humby et al. (2019)	144 early (<1 year) treatment naïve RA	US-guided synovial biopsy	sDMARDs	Biopsy at 0 and 6 months	IHC and nanostring	Patients with a lympho-myeloid pathotype have significantly higher disease severity, autoantibody positivity, and baseline erosive load. Myeloid- and lymphoid-associated gene expression strongly correlated with disease activity and acute phase reactants
Lliso-Ribera 2019	Lliso-Ribera et al. (2019)	200 early patients with inflammatory arthritis	US-guided synovial biopsy	sDMARDs	Biopsy at 0 and 6 months	IHC	Patients fulfilling the 1987 RA criteria had significantly higher levels of disease activity, histological synovitis, degree of immune cell infiltration, and differential upregulation of genes involved in B and T cell activation/function compared with RA 2010 criteria or UA, which shared similar clinical and pathobiological feature

more recent publication in early RA has further highlighted the value of synovial tissue analyses in refining the diagnosis of RA vs undifferentiated arthritis (Lliso-Ribera et al. 2019).

Moreover, deep phenotyping of synovial tissue by molecular analyses has identified specific gene signatures associated with clinical phenotype. In particular, for example, peripheral blood interferon response genes were associated with the lymhpo-myeloid pathotype, while synovial plasma cell signature was associated with progression of structural damage (Lewis et al. 2019). Additional analyses from the Accelerating Medicine Partnership (AMP) group, by integrating single cell RNA sequencing and mass cytometry, have recently identified unique cell population expanded in RA synovia that allow to distinguish the degree of synovial inflammation (Rao et al. 2017; Zhang et al. 2019). Specific cell populations included HY1 (CD90) +HLA-DRAhi sublining fibroblasts, IL1B+ pro-inflammatory monocytes, ITGAX + TBX21 + autoimmune-associated B cells, and PDCD1+ peripheral helper T (TPH) cells and follicular helper T (TFH) cells. The latter, in particular, are essential for the formation of TLO and have been already discussed in the previous paragraph. However, to date, little is known about the association of these cell types with disease features, such as disease severity, progression, and response to treatment. In the near future, it will be of utmost importance to confirm the relevance of these immune populations, by studying their association with clinical phenotype in larger cohorts of patients with RA.

3.3.2 TLO as Direct Therapeutic Targets

Because of their well-established relevance in driving the pathogenesis of RA and their association with worse disease outcomes, several strategies aiming at targeting TLO in RA have also been tested.

A number of studies have attempted to target mediators that are relevant in the formation or maintenance of TLO. The modulation of the IL-21/IL-21R pathway as a treatment strategy was first tested in experimental models of RA. IL-21R deficiency in the K/BxN mouse model of inflammatory arthritis (Kim et al. 2009) and antigen-induced arthritis (Roeleveld et al. 2017) is sufficient to block RA initiation, while the blockade of the IL-21/IL-21R pathway ameliorates disease in collagen-induced arthritis models treated with murine IL-21R Fc fusion protein (Young et al. 2007). However, there are still no data in patients with RA on the blockade of IL-21/IL-21R.

Some other molecules have been already tested in patients, but results have not been particularly striking, as in the case of inhibiting LTβ, which did not show clinical efficacy (Bienkowska et al. 2014). Similarly, drugs inhibiting IL-17 and IL-12/IL-23 showed little or no differences compared with placebo in RA (Kerschbaumer et al. 2019). This is in contrast with data on seronegative arthritis, where inhibition of IL-17 and its axis proved to be extremely effective, although it has been suggested that the analysis of targeted expression of these molecules could potentially help in predicting treatment response (Boutet et al. 2018).

Importantly, none of the above studies targeting mediators involved in TLO formation or maintenance in RA has stratified patients on the basis of TLO presence, which could have helped in selecting patients with higher chances of response.

3.3.3 TLO as Predictors of Treatment Response

As highlighted in the previous paragraphs, the presence of TLO is able to identify a subset of RA patients with a specific disease phenotype, specifically higher disease activity and higher prevalence of autoantibodies. Therefore, it is plausible to hypothesize that the presence of TLO could help to predict treatment response. A number of studies have explored the analyses of synovial tissues to predict treatment response. However, relatively few included the systematic analysis of TLO. Futhermore, because of the relatively small number of patients, the inconsistency in the definition of TLO, and the use of different time points for repeated biopsy, most of the results are fragmented and difficult to interpret.

Canete et al., for example, demonstrated significantly lower response in patients who were TLO positive despite a significantly higher use of anti-TNFa agents. (Cañete et al. 2009) By linear regression, TLO positive were found to predict lack of response to anti-TNFa. In this study, however, patients started sequential treatment with escalation to anti-TNFa in non-responders, and therefore, there could have been a selection of TLO + patients as the most severe, thus non-responders.

On the contrary, Klaasen et al., by analyzing synovial samples obtained before and after standardized treatment with infliximab in a cohort of 97 patients, found that the presence of TLO at baseline was a highly significant predictor of the clinical response to anti-TNF treatment (Klaasen et al. 2009).

More recently, Dennis et al. provided the molecular confirmation of the histological pathotypes previously described by histology. In addition, by analyzing the data from a previous cohort undergoing treatment with infliximab, they were able to identify TLO signature as predictor of response to TNFi (Dennis et al. 2014). The limitation of this manuscript consisted in the analysis of synovial samples obtained from arthroplasty, thus without standardization of treatment.

The observations published from our early RA cohort allowed to overcome such limitations and have shown a reduction of lymphoid-associated genes in EULAR good responders to csDMARDs (Humby et al. 2019). Similarly, molecular analyses by RNAseq identified a number of cell modules, including B cells, in association with B response to csDMARDs (Lewis et al. 2019). Importantly, these data come from the analysis of synovial tissue obtained by US-guided synovial biopsies in untreated patients with early Rheumatoid Arthritis, thus eliminating the bias of long-standing disease, treatment or the exclusive inclusion of large joints in studies based on arthroscopy.

In recent years, continuing on the same line, two international consortia have driven the delivery of the first two large-scale biopsy-driven RCTs in Rheumatoid Arthritis. As part of a study funded by the UK National Institute of Health

Research, a randomized, open labeled study in anti-TNFa inadequate responders to investigate the mechanisms for Response—resistance to rituximab versus tocilizumab in RA (R4-RA), a total of 165 patients failing treatment with TNFi have been recruited. Promising preliminary results were presented at the ACR 2019, while the trial is currently being analyzed and final results will be soon published. Similarly, as part of the MRC and Versus Arthritis-funded consortium MAximizing Therapeutic Utility in RA (MATURA), the Stratifying Therapies for Rheumatoid Arthritis by Pathobiology (STRAP) RCT has enrolled a total of 226 patients who failed csDMARDs and is due to being completed in the last quarter of 2020.

These studies have been appropriately powered and thus will hopefully give clear answers on the utility of synovial biopsy analysis in predicting treatment response in RA. Specifically, the studies aimed at understanding if patients lacking synovial B cells have a lower response to B cell targeted treatment (Rituximab) as opposed to other treatments. At the same time, the studies will provide invaluable information to answer additional research questions, including the association of TLO with disease severity, progression, and treatment response.

4 Conclusions

Here, we offered a comprehensive review on the relevance of synovial TLO in RA. The data presented indicate that the ontogeny of TLO resembles the development of secondary lymphoid organs, since many of the mediators known to be involved in lymphoneogenesis have been identified in the synotia of RA patients. Importantly, these structures are fully functional, as they induce the local maturation of B cells toward the production of autoantibodies. Their presence has been described in about 40% of patients with RA from early disease stages and has been strongly associated with disease severity and progression. Despite the availability of several drug treatments that can directly or indirectly target TLO and their components, a stratified medicine approach is needed to fully appreciate the potential effect of such treatments.

Bibliography

Amara K, Steen J, Murray F, Morbach H, Fernandez-Rodriguez BM, Joshua V et al (2013) Monoclonal IgG antibodies generated from joint-derived B cells of RA patients have a strong bias toward citrullinated autoantigen recognition. J Exp Med. https://doi.org/10.1084/jem.20121486

Baeten D, Peene I, Union A, Meheus L, Sebbag M, Serre G et al (2001) Specific presence of intracellular citrullinated proteins in rheumatoid arthritis synovium: Relevance to antifilaggrin autoantibodies. Arthritis Rheum. https://doi.org/10.1002/1529-0131(200110)44:10%3c2255:aid-art388%3e3.0.co;2-%23

Bar-Ephraim YE, Mebius RE (2016) Innate lymphoid cells in secondary lymphoid organs. Immunol Rev 185–99. https://doi.org/10.1111/imr.12407

Barone F, Nayar S, Campos J, Cloake T, Withers DR, Toellner K-M et al (2015) IL-22 regulates lymphoid chemokine production and assembly of tertiary lymphoid organs. Proc Natl Acad Sci 112:11024–11029. https://doi.org/10.1073/pnas.1503315112

Barone F, Gardner DH, Nayar S, Steinthal N, Buckley CD, Luther SA (2016) Stromal fibroblasts in tertiary lymphoid structures: a novel target in chronic inflammation. Front Immunol 7:477. https://doi.org/10.3389/fimmu.2016.00477

Bienkowska J, Norm A, Thai A, Goyal J, Plavina T, Nirula A et al (2014). Lymphotoxin-LIGHT pathway regulates the interferon signature in rheumatoid arthritis. PLoS One 9. https://doi.org/10.1371/journal.pone.0112545

Bombardieri M, Lewis M, Pitzalis C (2017) Ectopic lymphoid neogenesis in rheumatic autoimmune diseases. Nat Rev Rheumatol 13:141–154. https://doi.org/10.1038/nrrheum.2016.217

Botta D, Fuller MJ, Marquez-Lago TT, Bachus H, Bradley JE, Weinmann AS et al (2017) Dynamic regulation of T follicular regulatory cell responses by interleukin 2 during influenza infection. Nat Immunol. https://doi.org/10.1038/ni.3837

Boutet MA, Nerviani A, Gallo Afflitto G, Pitzalis C (2018) Role of the IL-23/IL-17 axis in psoriasis and psoriatic arthritis: The clinical importance of its divergence in skin and joints. Int J Mol Sci 19. https://doi.org/10.3390/ijms19020530

Bradfield PF, Amft N, Vernon-Wilson E, Exley AE, Parsonage G, Rainger GE et al (2003) Rheumatoid fibroblast-like synoviocytes overexpress the chemokine stromal cell-derived factor 1 (CXCL12), which supports distinct patterns and rates of $CD4^+$ and $CD8^+$ T cell migration within synovial tissue. Arthritis Rheum 48:2472–2482. https://doi.org/10.1002/art.11219

Bugatti S, Caporali R, Manzo A, Vitolo B, Pitzalis C, Montecucco C (2005) Involvement of subchondral bone marrow in rheumatoid arthritis: lymphoid neogenesis and in situ relationship to subchondral bone marrow osteoclast recruitment. Arthritis Rheum 52:3448–3459

Bugatti S, Manzo A, Vitolo B, Benaglio F, Binda E, Scarabelli M et al (2014) High expression levels of the B cell chemoattractant CXCL13 in rheumatoid synovium are a marker of severe disease. Rheumatol (United Kingdom) 53:1886–1895. https://doi.org/10.1093/rheumatology/keu163

Cañete JD, Celis R, Moll C, Izquierdo E, Marsal S, Sanmartí R et al (2009) Clinical significance of synovial lymphoid neogenesis and its reversal after anti-tumour necrosis factor alpha therapy in rheumatoid arthritis. Ann Rheum Dis 68:751–756

Cañete JD, Celis R, Yeremenko N, Sanmartí R, Van Duivenvoorde L, Ramírez J et al (2011) Ectopic lymphoid neogenesis is strongly associated with activation of the IL-23 pathway in rheumatoid synovitis. https://doi.org/10.1186/s13075-015-0688-0

Cantaert T, Kolln J, Timmer T, van der Pouw Kraan TC, Vandooren B, Thurlings RM et al (2008) B lymphocyte autoimmunity in rheumatoid synovitis is independent of ectopic lymphoid neogenesis. J Immunol 181:785–794. https://doi.org/10.4049/jimmunol.181.1.785

Chabaud M, Durand JM, Buchs N, Fossiez F, Page G, Frappart L et al (1999) Human interleukin-17: a T cell-derived proinflammatory cytokine produced by the rheumatoid synovium. Arthritis Rheum 42:963–970. https://doi.org/10.1002/1529-0131(199905)42:5%3c963:AID-ANR15%3e3.0.CO;2-E

Chu Y, Wang F, Zhou M, Chen L, Lu Y (2014) A preliminary study on the characterization of follicular helper T (Tfh) cells in rheumatoid arthritis synovium. Acta Histochem 116:539–543. https://doi.org/10.1016/j.acthis.2013.10.009

Corsiero E, Bombardieri M, Manzo A, Bugatti S, Uguccioni M, Pitzalis C (2012) Role of lymphoid chemokines in the development of functional ectopic lymphoid structures in rheumatic autoimmune diseases. Immunol Lett 145:62–67. https://doi.org/10.1016/j.imlet.2012.04.013

Corsiero E, Bombardieri M, Carlotti E, Pratesi F, Robinson W, Migliorini P et al (2016) Single cell cloning and recombinant monoclonal antibodies generation from RA synovial B cells reveal frequent targeting of citrullinated histones of NETs. Ann Rheum Dis 75:1866–1875. https://doi.org/10.1136/annrheumdis-2015-208356

Corsiero E, Jagemann L, Perretti M, Pitzalis C, Bombardieri M (2018) Characterization of a synovial B cell-derived recombinant monoclonal antibody targeting stromal calreticulin in the rheumatoid joints. J Immunol 201:1373–1381. https://doi.org/10.4049/jimmunol.1800346

de Hair MJH, van de Sande MGH, Ramwadhdoebe TH, Hansson M, Landewé R, van der Leij C et al (2013) Features of the synovium of individuals at risk of developing rheumatoid arthritis: implications for understanding preclinical rheumatoid arthritis. Arthritis Rheum 66:513–522. https://doi.org/10.1002/art.38273

De Rycke L, Nicholas AP, Cantaert T, Kruithof E, Echols JD, Vandekerckhove B et al (2005) Synovial intracellular citrullinated proteins colocalizing with peptidyl arginine deiminase as pathophysiologically relevant antigenic determinants of rheumatoid arthritis-specific humoral autoimmunity. Arthritis Rheum. https://doi.org/10.1002/art.21220

De Silva NS, Klein U (2015) Dynamics of B cells in germinal centres. Nat Rev Immunol 15:137–148. https://doi.org/10.1038/nri3804

Dennis G, Holweg CTJ, Kummerfeld SK, Choy DF, Setiadi AF, Hackney JA et al (2014) Synovial phenotypes in rheumatoid arthritis correlate with response to biologic therapeutics. Arthritis Res Ther 16:R90. https://doi.org/10.1186/ar4555

Derksen VFAM, Huizinga TWJ, van der Woude D (2017) The role of autoantibodies in the pathophysiology of rheumatoid arthritis. Semin Immunopathol 39:437–446. https://doi.org/10.1007/s00281-017-0627-z

Ding Y, Li J, Yang PA, Luo B, Wu Q, Zajac AJ et al (2014) Interleukin-21 promotes germinal center reaction by skewing the follicular regulatory T cell to follicular helper T cell balance in autoimmune BXD2 mice. Arthritis Rheumatol. https://doi.org/10.1002/art.38735

Drayton DL, Liao S, Mounzer RH, Ruddle NH (2006) Lymphoid organ development: from ontogeny to neogenesis. Nat Immunol 7:344–353. https://doi.org/10.1038/ni1330

Fonseca VR, Agua-Doce A, Maceiras AR, Pierson W, Ribeiro F, Romão VC et al (2017) Human blood Tfr cells are indicators of ongoing humoral activity not fully licensed with suppressive function. Sci Immunol. https://doi.org/10.1126/sciimmunol.aan1487

Gensous N, Charrier M, Duluc D, Contin-Bordes C, Truchetet ME, Lazaro E et al (2018) T follicular helper cells in autoimmune disorders. Front Immunol 9:1637. https://doi.org/10.3389/fimmu.2018.01637

Germar K, Fehres CM, Scherer HU, van Uden N, Pollastro S, Yeremenko N et al (2019) Generation and characterization of anti-citrullinated protein antibody-producing B cell clones from rheumatoid arthritis patients. Arthritis Rheum. https://doi.org/10.1002/art.40739

Gómez-Puerta JA, Celis R, Hernández MV, Ruiz-Esquide V, Ramírez J, Haro I et al (2013) Differences in synovial fluid cytokine levels but not in synovial tissue cell infiltrate between anti-citrullinated peptide/protein antibody-positive and -negative rheumatoid arthritis patients. Arthritis Res Ther 15:R182. https://doi.org/10.1186/ar4372

He J, Tsai LM, Leong YA, Hu X, Ma CS, Chevalier N et al (2013) Circulating precursor CCR7loPD-1hi $CXCR5^+$ $CD4^+$ T cells indicate Tfh cell activity and promote antibody responses upon antigen reexposure. Immunity 39:770–781. https://doi.org/10.1016/j.immuni.2013.09.007

Humby F, Bombardieri M, Manzo A, Kelly S, Blades MC, Kirkham B et al (2009) Ectopic lymphoid structures support ongoing production of class-switched autoantibodies in rheumatoid synovium. PLoS Med 6:e1. https://doi.org/10.1371/journal.pmed.0060001

Humby F, Kelly S, Bugatti S, Manzo A, Filer A, Mahto A et al (2016) Evaluation of minimally invasive, ultrasound-guided synovial biopsy techniques by the OMERACT filter—determining validation requirements. J Rheumatol 43:208–213. https://doi.org/10.3899/jrheum.141199

Humby F, Lewis M, Ramamoorthi N, Hackney JA, Barnes MR, Bombardieri M et al (2019) Synovial cellular and molecular signatures stratify clinical response to csDMARD therapy and predict radiographic progression in early rheumatoid arthritis patients. Ann Rheum Dis 78:761–772. https://doi.org/10.1136/annrheumdis-2018-214539

Ikeuchi H, Kuroiwa T, Hiramatsu N, Kaneko Y, Hiromura K, Ueki K et al (2005) Expression of interleukin-22 in rheumatoid arthritis: potential role as a proinflammatory cytokine. Arthritis Rheum 52:1037–1046. https://doi.org/10.1002/art.20965

Jones GW, Jones SA (2016) Ectopic lymphoid follicles: inducible centres for generating antigen-specific immune responses within tissues. Immunology 147:141–151. https://doi.org/10.1111/imm.12554

Jones GW, Bombardieri M, Greenhill CJ, McLeod L, Nerviani A, Rocher-Ros V et al (2015) Interleukin-27 inhibits ectopic lymphoid-like structure development in early inflammatory arthritis. J Exp Med 212:1793–1802. https://doi.org/10.1084/jem.20132307

Jüngel A, Distler JHW, Kurowska-Stolarska M, Seemayer CA, Seibl R, Forster A et al (2004) Expression of interleukin-21 receptor, but not interleukin-21, in synovial fibroblasts and synovial macrophages of patients with rheumatoid arthritis. Arthritis Rheum 50:1468–1476. https://doi.org/10.1002/art.20218

Karnell JL, Ettinger R (2012) The interplay of IL-21 and BAFF in the formation and maintenance of human B cell memory. Front Immunol 3. https://doi.org/10.3389/fimmu.2012.00002

Kerschbaumer A, Sepriano A, Smolen JS, van der Heijde D, Dougados M, van Vollenhoven R et al (2019) Efficacy of pharmacological treatment in rheumatoid arthritis: a systematic literature research informing the, update of the EULAR recommendations for management of rheumatoid arthritis. Ann Rheum Dis 2020:1–15. https://doi.org/10.1136/annrheumdis-2019-216656

Kim JM, Youn J, Jang E, Cho S-H, Park H, Paik D-J (2009) K/BxN mice development of arthritis in autoimmune cell expansion is essential for the T−CD25+ provoked by homeostatic CD4 a positive feedback loop of IL-21 signaling. J Immunol Ref 182:4649–4656. https://doi.org/10.4049/jimmunol.0804350

Klaasen R, Thurlings RM, Wijbrandts CA, van Kuijk AW, Baeten D, Gerlag DM et al (2009) The relationship between synovial lymphocyte aggregates and the clinical response to infliximab in rheumatoid arthritis: a prospective study. Arthritis Rheum 60:3217–3224

Klimiuk PA, Goronzy JJ, Björnsson J, Beckenbaugh RD, Weyand CM (1997) Tissue cytokine patterns distinguish variants of rheumatoid synovitis. Am J Pathol 151:1311–1319

Kotake S, Udagawa N, Hakoda M, Mogi M, Yano K, Tsuda E et al (2001) Activated human T cells directly induce osteoclastogenesis from human monocytes: possible role of T cells in bone destruction in rheumatoid arthritis patients. Arthritis Rheum 44:1003–1012. https://doi.org/10.1002/1529-0131(200105)44:5%3c1003:aid-anr179%3e3.0.co;2-%23

Kwok SK, La Cho M, Park MK, Oh HJ, Park JS, Her YM et al (2012) Interleukin-21 promotes osteoclastogenesis in humans with rheumatoid arthritis and in mice with collagen-induced arthritis. Arthritis Rheum 64:740–751. https://doi.org/10.1002/art.33390

Lee SY, Jung YO, Ryu JG, Kang CM, Kim EK, Son HJ et al (2014) Intravenous immunoglobulin attenuates experimental autoimmune arthritis by inducing reciprocal regulation of th17 and treg cells in an interleukin-10- dependent manner. Arthritis Rheum. https://doi.org/10.1002/art.38627

Lewis MJ, Barnes MR, Blighe K, Goldmann K, Rana S, Hackney JA et al (2019) Molecular portraits of early rheumatoid arthritis identify clinical and treatment response phenotypes. Cell Rep 28(2455–2470):e5. https://doi.org/10.1016/j.celrep.2019.07.091

Linn-Rasker SP, Van Der Helm-Van Mil a. HM, Breedveld FC, Huizinga TWJ (2007) Arthritis of the large joints—in particular, the knee—at first presentation is predictive for a high level of radiological destruction of the small joints in rheumatoid arthritis. Ann Rheum Dis 66:646–650. https://doi.org/10.1136/ard.2006.066704

Linterman MA, Pierson W, Lee SK, Kallies A, Kawamoto S, Rayner TF et al (2011) Foxp3 + follicular regulatory T cells control the germinal center response. Nat Med 17:975–982. https://doi.org/10.1038/nm.2425

Liu D, Xu H, Shih C, Wan Z, Ma X, Ma W et al (2015) T-B-cell entanglement and ICOSL-driven feed-forward regulation of germinal centre reaction. Nature 517:214–218. https://doi.org/10.1038/nature13803

Liu C, Wang D, Lu S, Xu Q, Zhao L, Zhao J et al (2018) Increased circulating follicular Treg cells are associated with lower levels of auto antibodies in patients with rheumatoid arthritis in stable remission. Arthritis Rheum. https://doi.org/10.1002/art.40430

Lliso-Ribera G, Humby F, Lewis M (n.d.) Synovial tissue signatures enhance clinical classification and prognostic/treatment response algorithms in early inflammatory arthritis and predict requirement for subsequent biologic therapy: results from the Pathobiology of Early Arthritis Cohort (PEAC). Ann Rheum Dis https://doi.org/10.1136/annrheumdis-2019-215751, https://doi.org/10.1136/annrheumdis-2019-215751

Lliso-Ribera G, Humby F, Lewis M, Nerviani A, Mauro D, Rivellese F et al (2019) Synovial tissue signatures enhance clinical classification and prognostic/treatment response algorithms in early inflammatory arthritis and predict requirement for subsequent biological therapy: results from the pathobiology of early arthritis cohort (PEA. Ann Rheum Dis 78:1642–1652. https://doi.org/10.1136/annrheumdis-2019-215751

Luther SA, Lopez T, Bai W, Hanahan D, Cyster JG (2000) BLC expression in pancreatic islets causes B cell recruitment and lymphotoxin-dependent lymphoid neogenesis. Immunity 12:471–481. https://doi.org/10.1016/s1074-7613(00)80199-5

Ma J, Zhu C, Ma B, Tian J, Baidoo SE, Mao C et al (2012) Increased frequency of circulating follicular helper T cells in patients with rheumatoid arthritis. Clin Dev Immunol 2012:827480. https://doi.org/10.1155/2012/827480

Manzo A, Paoletti S, Carulli M, Blades MC, Barone F, Yanni G et al (2005) Systematic microanatomical analysis of CXCL13 and CCL21 in situ production and progressive lymphoid organization in rheumatoid synovitis. Eur J Immunol 35:1347–1359. https://doi.org/10.1002/eji.200425830

Manzo A, Bugatti S, Caporali R, Prevo R, Jackson DG, Uguccioni M et al (2007) CCL21 expression pattern of human secondary lymphoid organ stroma is conserved in inflammatory lesions with lymphoid neogenesis. Am J Pathol. https://doi.org/10.2353/ajpath.2007.061275

Manzo A, Vitolo B, Humby F, Caporali R, Jarrossay D, Dell'accio F et al (2008) Mature antigen-experienced T helper cells synthesize and secrete the B cell chemoattractant CXCL13 in the inflammatory environment of the rheumatoid joint. Arthritis Rheum 58:3377–3387. https://doi.org/10.1002/art.23966

Masson-Bessière C, Sebbag M, Durieux JJ, Nogueira L, Vincent C, Girbal-Neuhauser E et al (2000) In the rheumatoid pannus, anti-filaggrin autoantibodies are produced by local plasma cells and constitute a higher proportion of IgG than in synovial fluid and serum. Clin Exp Immunol. https://doi.org/10.1046/j.1365-2249.2000.01171.x

McCausland MM, Yusuf I, Tran H, Ono N, Yanagi Y, Crotty S (2014) SAP regulation of follicular helper CD4 T cell development and humoral immunity is independent of SLAM and Fyn Kinase. J Immunol 178:817–828. https://doi.org/10.4049/jimmunol.178.2.817

McInnes IB, Schett G (2011) The pathogenesis of rheumatoid arthritis. N Engl J Med 365:2205–2219. https://doi.org/10.7748/phc2011.11.21.9.29.c8797

McInnes IB, Schett G (2017) Pathogenetic insights from the treatment of rheumatoid arthritis. Lancet 389:2328–2337. https://doi.org/10.1016/S0140-6736(17)31472-1

Morita R, Schmitt N, Bentebibel SE, Ranganathan R, Bourdery L, Zurawski G et al (2011) Human blood CXCR5+ CD4+ T cells are counterparts of T follicular cells and contain specific subsets that differentially support antibody secretion. Immunity 34:108–121. https://doi.org/10.1016/j.immuni.2010.12.012

Moschovakis GL, Bubke A, Friedrichsen M, Falk CS, Feederle R, Förster R (2017) T cell specific Cxcr5 deficiency prevents rheumatoid arthritis. Sci Rep 7. https://doi.org/10.1038/s41598-017-08935-6

Nayar S, Campos J, Smith CG, Iannizzotto V, Gardner DH, Mourcin F et al (2019) Immunofibroblasts are pivotal drivers of tertiary lymphoid structure formation and local pathology. Proc Natl Acad Sci 116:13490–13497. https://doi.org/10.1073/pnas.1905301116

Niu Q, Huang Z, Wu X, Jin Y, An Y, Li Y et al (2018) Enhanced IL-6/phosphorylated STAT3 signaling is related to the imbalance of circulating T follicular helper/T follicular regulatory cells in patients with rheumatoid arthritis. Arthritis Res Ther. https://doi.org/10.1186/s13075-018-1690-0

Orr C, Najm A, Biniecka M, McGarry T, Ng C-T, Young F et al (2017) Synovial immunophenotype and anti-citrullinated peptide antibodies in rheumatoid arthritis patients: relationship to treatment response and radiologic prognosis. Arthritis Rheum (Hoboken, NJ) 69:2114–2123. https://doi.org/10.1002/art.40218

Penatti A, Facciotti F, De Matteis R, Larghi P, Paroni M, Murgo A et al (2017) Differences in serum and synovial CD4+ T cells and cytokine profiles to stratify patients with inflammatory osteoarthritis and rheumatoid arthritis. Arthritis Res Ther 19:103. https://doi.org/10.1186/s13075-017-1305-1

Pitzalis C, Kelly S, Humby F (2013) New learnings on the pathophysiology of RA from synovial biopsies. Curr Opin Rheumatol 25:334–344. https://doi.org/10.1097/BOR.0b013e32835fd8eb

Randall TD, Carragher DM, Rangel-Moreno J (2008) Development of Secondary lymphoid organs. Annu Rev Immunol 26:627–650. https://doi.org/10.1146/annurev.immunol.26.021607.090257

Rao DA, Gurish MF, Marshall JL, Slowikowski K, Fonseka CY, Liu Y et al (2017) Pathologically expanded peripheral T helper cell subset drives B cells in rheumatoid arthritis. Nat Publ Gr 542:110–114. https://doi.org/10.1038/nature20810

Ren J, Feng Z, Lv Z, Chen X, Li J (2011) Natural killer-22 cells in the synovial fluid of patients with rheumatoid arthritis are an innate source of interleukin 22 and tumor necrosis factor-α. J Rheumatol 38:2112–2118. https://doi.org/10.3899/jrheum.101377

Rivellese F, Nerviani A, Rossi FW, Marone G, Matucci-Cerinic M, de Paulis A et al (2017) Mast cells in rheumatoid arthritis: friends or foes? Autoimmun Rev 16:557–563. https://doi.org/10.1016/j.autrev.2017.04.001

Rivellese F, Mauro D, Nerviani A, Pagani S, Fossati-Jimack L, Messemaker T et al (2018) Mast cells in early rheumatoid arthritis associate with disease severity and support B cell autoantibody production. Ann Rheum Dis 77:1773–1781

Rivellese F, Humby F, Bugatti S, Fossati-Jimack L, Rizvi H, Lucchesi D et al (2019a) B cell synovitis and clinical phenotypes in rheumatoid arthritis: relationship to disease stages and drug exposure. Arthritis Rheum 4–11. https://doi.org/10.1002/art.41184

Rivellese F, Rossi FW, Galdiero MR, Pitzalis C, de Paulis A (2019b) Mast cells in early rheumatoid arthritis. Int J Mol Sci 20:1–13. https://doi.org/10.3390/ijms20082040

Roeleveld DM, Marijnissen RJ, Walgreen B, Helsen MM, Van Den Bersselaar L, Van De Loo FA et al (2017) Higher efficacy of anti-IL-6/IL-21 combination therapy compared to monotherapy in the induction phase of Th17-driven experimental arthritis 2017. https://doi.org/10.1371/journal.pone.0171757

Romão VC, Fonseca JE, Agua-Doce A, Graca L (2018) T follicular regulatory cells are decreased in patients with established treated rheumatoid arthritis with active disease: comment on the article by Liu et al. Arthritis Rheum. https://doi.org/10.1002/art.40586

Ronday HK, Van Der Laan WH, Tak PP, De Roos JADM, Bank RA, Tekoppele JM et al (n.d.) Human granzyme B mediates cartilage proteoglycan degradation and is expressed at the invasive front of the synovium in rheumatoid arthritis

Sawa S, Cherrier M, Lochner M, Satoh-Takayama N, Fehling HJ, Langa F et al (2010) Lineage relationship analysis of RORγt+ innate lymphoid cells. Science (80-) 330:665–669. https://doi.org/10.1126/science.1194597

Scheel T, Gursche A, Zacher J, Berek C (2011) V-region gene analysis of locally defined synovial B and plasma cells reveals selected B cell expansion and accumulation of plasma cell clones in rheumatoid arthritis. Arthritis Rheum. https://doi.org/10.1002/art.27767

Smolen JS, Aletaha D, McInnes IB (2016) Rheumatoid arthritis. Lancet 388:2023–2038. https://doi.org/10.1016/S0140-6736(16)30173-8

Stumhofer JS, Laurence A, Wilson EH, Huang E, Tato CM, Johnson LM et al (2006) Interleukin 27 negatively regulates the development of interleukin 17-producing T helper cells during chronic inflammation of the central nervous system. Nat Immunol. https://doi.org/10.1038/ni1376

Tak PP, Kummer JA, Hack CE, Daha MR, Smeets TJM, Erkelens GW et al (1994) Granzyme-positive cytotoxic cells are specifically increased in early rheumatoid synovial tissue. Arthritis Rheum 37:1735–1743. https://doi.org/10.1002/art.1780371205

Takemura S, Braun A, Crowson C, Kurtin PJ, Cofield RH, O'Fallon WM et al (2001) Lymphoid neogenesis in rheumatoid synovitis. J Immunol. https://doi.org/10.4049/jimmunol.167.2.1072

Thurlings RM, Wijbrandts CA, Mebius RE, Cantaert T, Dinant HJ, Van Der Pouw-Kraan TCTMTM et al (2008) Synovial lymphoid neogenesis does not define a specific clinical rheumatoid arthritis phenotype. Arthritis Rheum 58:1582–1589. https://doi.org/10.1002/art.23505

Timmer TCG, Baltus B, Vondenhoff M, Huizinga TWJ, Tak PP, Verweij CL et al (2007) Inflammation and ectopic lymphoid structures in rheumatoid arthritis synovial tissues dissected by genomics technology: identification of the interleukin-7 signaling pathway in tissues with lymphoid neogenesis. Arthritis Rheum 56:2492–2502. https://doi.org/10.1002/art.22748

Trouw LA, Rispens T, Toes REM (2017) Beyond citrullination: other post-translational protein modifications in rheumatoid arthritis. Nat Rev Rheumatol. https://doi.org/10.1038/nrrheum.2017.15

Tsai LM, Yu D (2014) Follicular helper T-cell memory: establishing new frontiers during antibody response. Immunol Cell Biol 92:57–63. https://doi.org/10.1038/icb.2013.68

Ueno H, Banchereau J, Vinuesa CG (2015) Pathophysiology of T follicular helper cells in humans and mice. Nat Immunol 16:142–152. https://doi.org/10.1038/ni.3054

Van De Sande MGHH, Thurlings RM, Boumans MJHH, Wijbrandts CA, Modesti MG, Gerlag DM et al (2011) Presence of lymphocyte aggregates in the synovium of patients with early arthritis in relationship to diagnosis and outcome: is it a constant feature over time? Ann Rheum Dis 70:700–703. https://doi.org/10.1136/ard.2010.139287

van Delft MAM, Huizinga TWJ (2020) An overview of autoantibodies in rheumatoid arthritis. J Autoimmun. https://doi.org/10.1016/j.jaut.2019.102392

Van Oosterhout M, Bajema I, Levarht EWN, Toes REM, Huizinga TWJ, Van Laar JM (2008) Differences in synovial tissue infiltrates between anti-cyclic citrullinated peptide-positive rheumatoid arthritis and anti-cyclic citrullinated peptide-negative rheumatoid arthritis. Arthritis Rheum 58:53–60. https://doi.org/10.1002/art.23148

Vossenaar ER, Smeets TJM, Kraan MC, Raats JM, Van Venrooij WJ, Tak PP (2004) The presence of citrullinated proteins is not specific for rheumatoid synovial tissue. Arthritis Rheum. https://doi.org/10.1002/art.20584

Wang X, Yang C, Xu F, Qi L, Wang J, Yang P (2019) Imbalance of circulating Tfr/Tfh ratio in patients with rheumatoid arthritis. Clin Exp Med. https://doi.org/10.1007/s10238-018-0530-5

Willemze A, Trouw LA, Toes REM, Huizinga TWJ (2012) The influence of ACPA status and characteristics on the course of RA. Nat Rev Rheumatol 8:144–152. https://doi.org/10.1038/nrrheum.2011.204

Wing JB, Ise W, Kurosaki T, Sakaguchi S (2014) Regulatory T cells control antigen-specific expansion of Tfh cell number and humoral immune responses via the coreceptor CTLA-4. Immunity. https://doi.org/10.1016/j.immuni.2014.12.006

Wu H, Chen Y, Liu H, Xu LL, Teuscher P, Wang S et al (2016) Follicular regulatory T cells repress cytokine production by follicular helper T cells and optimize IgG responses in mice. Eur J Immunol. https://doi.org/10.1002/eji.201546094

Xing R, Jin Y, Sun L, Yang L, Li C, Li Z et al (2016) Interleukin-21 induces migration and invasion of fibroblast-like synoviocytes from patients with rheumatoid arthritis. Clin Exp Immunol 184:147–158. https://doi.org/10.1111/cei.12751

Yamin R, Berhani O, Peleg H, Aamar S, Stein N, Gamliel M et al (2019) High percentages and activity of synovial fluid NK cells present in patients with advanced stage active rheumatoid arthritis. Sci Rep 9:1–12. https://doi.org/10.1038/s41598-018-37448-z

Yoshida H, Hunter CA (2015) The immunobiology of interleukin-27. Annu Rev Immunol. https://doi.org/10.1146/annurev-immunol-032414-112134

Young DA, Hegen M, Ma HLR, Whitters MJ, Albert LM, Lowe L et al (2007) Blockade of the interleukin-21/interleukin-21 receptor pathway ameliorates disease in animal models of rheumatoid arthritis. Arthritis Rheum 56:1152–1163. https://doi.org/10.1002/art.22452

Zhang F, Wei K, Slowikowski K, Fonseka CY, Rao DA, Kelly S et al (2019) Defining inflammatory cell states in rheumatoid arthritis joint synovial tissues by integrating single-cell transcriptomics and mass cytometry. Nat Immunol. https://doi.org/10.1038/s41590-019-0378-1

Zhu J, Jia E, Zhou Y, Xu J, Feng Z, Wang H et al (2015) Interleukin-22 secreted by NKp44 + natural killer cells promotes proliferation of fibroblast-like synoviocytes in rheumatoid arthritis. Medicine (Baltimore) 94:e2137. https://doi.org/10.1097/MD.0000000000002137

Artificial Construction of Immune Tissues/Organoids and Their Application for Immunological Intervention

Yuka Kobayashi and Takeshi Watanabe

Contents

Abstract Human-type lymphoid tissue organoids, which stably function in our body for a certain period of time or longer, may have a great potential as immune-stimulatory or immune-regulatory devices and could be utilized in the future for the treatment of various diseases such as cancer, severe infection, autoimmunity and congenital as well as acquired immunodeficiency resulting from severe infections or aging. In this review, we discuss about rationality and trials of the synthesis of immunologically functional lymphoid tissue organoids mainly in mouse. We have been recently trying to construct immunologically functioning human-type organoids, and the efforts are also briefly described.

Y. Kobayashi · T. Watanabe (✉)
Institute for Frontier Life and Medical Sciences, Kyoto University, Kyoto, Japan
e-mail: tthta_w@icloud.com

Y. Kobayashi
e-mail: yukacoba@infront.kyoto-u.ac.jp

Current Topics in Microbiology and Immunology (2020) 426: 143–160
https://doi.org/10.1007/82_2020_215

Published Online: 02 June 2020

1 Introduction

The various types of immune organs/tissues are classified based on their ontogeny, location and function. There are the primary lymphoid organs (PLOs: bone marrow and thymus in mammals, bursa of Fabricius in chicken, appendix in rabbit, etc.), the secondary lymphoid organs (SLOs: peripheral lymph nodes, spleen, tonsil, and gut-associated lymphatic tissues (GALT), etc.) and the tertiary lymphoid organs/structures (TLOs/TLSs). Immune cells, such as T cells, B cells and dendritic cells, are produced and undergo differentiation in the PLOs, whereas the immune cells initiate immune response to antigen stimulation in the SLOs. The PLOs and SLOs develop mostly during the fetal period. On the other hand, the tertiary lymphoid organs/structures (TLOs) are formed after birth at ectopic sites, often in response to physiological insults such as in inflammation, infection, autoimmune disease, allograft rejection and cancer. We previously reported the construction as well as the immune function of "artificially-constructed peripheral lymphoid tissues (aLTs)/organoids" in mice. Because aLTs are constructed after birth at ectopic sites such as in the renal subcapsular space, they should technically also be classified as TLOs. In this chapter, we discuss first the structure and function of TLOs including the aLTs. Then, we discuss the significance of TLOs and aLTs in maintenance of immune homeostasis, in particular, their role in conferring antitumor immunity. It is expected that human-type aLTs/organoids may have great potential as immune-stimulatory or -regulatory devices and could be utilized in the future for treatment of human cancer, severe infection and congenital as well as acquired immunodeficiency resulting from severe infections or aging. We have been recently trying to construct such human-type aLTs/organoids, and our efforts are also briefly described in this chapter.

2 TLO Formation

The formation of TLOs occurs postnatally or during adulthood, mainly at ectopic sites, such as inflammatory sites in the peripheral tissues/organs and in cancer tissues. In contrast to secondary lymphoid tissues, which are organized in the absence of antigenic stimulation, TLOs appear to be induced mostly in the presence of various antigens such as tumor antigens, allo-antigens and auto-antigens, or mitogens, in addition to the various inflammatory reactions. TLOs and SLOs have a similar fundamental structure, and the process of TLO formation recapitulates that of SLOs (Randall et al. 2008; Fu and Chaplin 1999; Mebius 2003; Thaunat et al. 2010a; Carragher et al. 2008; Koning and Mebius 2012; Drayton et al. 2006). It also depends on the presence of lymphoid tissue organizer (LTo) cells, also known as lymphoid tissue-specific stromal cells, which ectopically appear and are capable of forming lymphoid tissues at inflammatory sites or tumor tissues (Buckley et al. 2015; Turley et al. 2015; Johasson et al. 2016). At such sites, LTo like cells, which express lymphotoxin beta receptor (LTβR) and/or tumor necrosis factor alpha receptor (TNFαR) on their cell surface, are derived possibly from the local mesenchymal cells, fibroblasts, blood vessel or lymphatic endothelial cells or smooth muscle cells (Roozendaal and Mebius 2011; Fletcher et al. 2011; Bar-Ephraim et al. 2016; Chai et al. 2013; Ruddle 2014; Engelhard et al. 2018; Tschenig et al. 2018). These tentative LTo stromal cells can be activated through LTβR by its ligand lymphotoxin-alpha 1/beta 2 (LTα1β2), which is expressed on the bone marrow-derived lymphoid tissue inducer cells (LTi cells), or by its alternate ligand, LIGHT, also known as a tumor necrosis factor superfamily member 14 (TNFSF14), expressed on activated T cells (Gommerman and Browning 2003; Scheu et al. 2002; Wang et al. 2009). The LTo stromal cells are also activated by soluble ligands such as LTα or TNFα through TNFαreceptor. The activated tentative LTo stromal cells secrete various lymphocyte-tropic chemokines such as CCL19, CCL20, CXCL12 and CXCL13 and express various adhesion molecules such as VCAM1, ICAM1 and MadCAM1 on the cell surface. The ectopic expression of these chemokines and cell-adhesive factors recruits immune cells such as T cells, B cells and dendritic cells onto the stromal cells. Similar to SLOs, TLOs are mostly composed of a T cell region, supported by distinct stromal cells, the fibroblastic reticular cells (FRCs) and of a B cell region, supported by its own distinct stromal cells, the follicular dendritic cells (FDCs). Upon LTβR stimulation, FRCs and FDCs start to secrete chemokines that induce the accumulation of T cells and B cells, respectively. The formation of TLOs is also accompanied by the formation of high endothelial venules (HEVs) and lymphoid tissue-specific capillaries (Ager 2017; Ruddle 2014). Upon stimulation of LTβR expressed on HEVs with lymphotoxin (LTα1β2) or LIGHT (TNFSF-14), HEV-specific cubic structures are formed (Scheu et al. 2002; Wang et al. 2009; Johansson-Percival et al. 2015). HEVs are structurally distinct blood vessels that develop in all secondary lymphoid organs except the spleen. They express PNAd, which is a ligand for L selectin, and accelerate the recruitment of lymphocytes by secretion of chemokines. HEVs are critical for initiating and maintaining immune responses because they extract naïve and memory lymphocytes from the bloodstream. HEVs also develop in TLOs in cases of allografts and cancers. HEV neogenesis in TLOs is thought to facilitate the generation of tissue-destroying lymphocytes inside chronically inflamed tissues and cancers by promoting the migration of lymphocytes from the bloodstream into the lymphoid tissues (Jones et al. 2018). Follicular helper T (Tfh) cells are major components of the humoral immune response due to their pivotal role in germinal center (GC) formation and antibody affinity maturation following B cell isotype switching in the GC (46,89). Tfh cells in the TLO are often identified within sites of inflammation as well as in the microenvironment of various cancers, indicating that GC formation and affinity maturation of antibodies may occur also in TLOs (Salomonsson et al. 2003; Couillault et al. 2018; Fonseca et al. 2018; Silina et al. 2018).

3 TLOs for Immune Responses and Homeostasis Maintenance

TLOs often function as powerhouses of disease immunity (Roco et al. 2019). The typical sites at which the TLOs are formed include autoimmune disease sites, such as the synovial membrane in rheumatoid arthritis (Shi et al. 2001; Takemura et al. 2001; Young et al. 1984; Kobayashi et al. 2013), the aorta adventitia in aortic atherosclerosis (Grabner et al. 2009; Lotzer et al. 2010; Hu et al. 2015), the salivary gland in Sjögren syndrome (Salomonsson et al. 2003; Stott et al. 1998; Fonseca et al. 2018), the thymus in myasthenia gravis (Thomas et al. 1982; Weiss et al. 2013), the kidney pelvic well in lupus nephritis (Tschenig et al. 2018), the inner bronchial wall in asthma (de Leur et al. 2018), the renal allograft in case of acute rejection (Lin et al. 2019) and so on. TLO formation also occurs at infection sites (Neyt et al. 2012) or within and around cancer tissues (Dieu-Nosjean et al. 2014; Joyce and Fearon 2015; Hiraoka et al. 2016; Yu et al. 2007). Furthermore, TLO formation often occurs at graft rejection sites in transplanted organs (Nasr et al. 2007; Thaunat et al. 2010b; Sicard et al. 2016), which suggests that TLO might be also involved in regulating graft rejection.

For a long time, studies have mostly focused on the negative features of TLOs, such as their involvement in the induction and progression of inflammation or their potential as autoantibody production sites. Particularly in the case of autoimmune diseases like systemic lupus erythematosus (SLE), TLOs are considered to be possible sites of autoantibody diversification, which leads to pathological deterioration in these diseases (Shi et al. 2001; Takemura et al. 2001; Young et al. 1984; Kobayashi et al. 2013; Salomonsson et al. 2003; Stott et al. 1998; Thomas et al. 1982; Weiss et al. 2013; Neyt et al. 2012). However, recent studies have reported some beneficial functions of TLOs, such as protection against infections (Carragher et al. 2008; Halle et al. 2009; Hughes et al. 2016; Jones and Jones 2016), by inducing an antigen-specific immune response, antitumor activity as described below and regulation of inflammatory responses (Carragher et al. 2008). Thus, the role of TLOs in delaying disease progression is gradually gaining attention. In cases of patients with aortic sclerosis, LTβR is expressed by the vascular aortic endothelium. It has been reported that the formation of TLOs in the aorta adventitia through LTβR signaling inhibited the formation of aortic plaque, suggesting that the formation of TLOs may protect from aortic disease progression (Kobayashi et al. 2013; Grabner et al. 2009; Lotzer et al. 2010). Thus, these studies suggest that TLOs may contribute to the maintenance of immune homeostasis.

4 TLO Formation and Anti-cancer Activity

Immune responses against tumors are typically observed upon infiltration of immune cells, especially killer T cells (tumor tissue-infiltrating lymphocytes, TILs), into cancer tissues, and the TILs have been expected to contribute to cancer suppression (Crotty 2011; Rosenberg and Dudley 1990). It has been reported that the killer T cells can also be activated within the tumor tissues (Yu et al. 2004, 2007). It has been reported that TLO formation in tumors might give a favorable prognosis in various types of human cancers (Hiraoka et al. 2016; Dieu-Nosjean et al. 2016; Sautes-Fridman et al. 2019; Figenschau et al. 2015; Engelhard et al. 2018; Kuwabara et al. 2019). Concerning the suppressive effectiveness of TLOs against cancer, several factors have been suggested. Those include the dependency on the stage of the cancer (Colbeck et al. 2017), the extent of infiltrating CD8+ T lymphocytes and regulatory T cells in TLOs (Kuwabara et al. 2019; de Leur et al. 2018), the appearance of tumor-infiltrating plasma cells (Solinas et al. 2017) or B cells (Lin et al. 2019), the extent of checkpoint molecule expression (Weinstein et al. 2019), the structure or location of TLOs in relation to the tumor, origin or location of the tumor (Engelhard et al. 2018), the cytokine environment surrounding cancer (Dorraji et al. 2018), etc. Moreover, preoperative treatments may influence the efficiency of the TLOs. Recently, it has been reported that preoperative treatment with neoadjuvant chemotherapy (NAC) strongly enhanced the anti-cancer activity of TLOs within the tumor microenvironment (Kroeger et al. 2016). The NAC-treated group of pancreatic ductal adenocarcinoma patients with extremely poor prognoses demonstrated a more favorable outcome compared to NAC-nontreated patients with TLOs. NAC induced a significantly higher proportion of CD8+ T cells, PNAd+ high endothelial venules (HEVs), CD63+ macrophages and Ki-67+ cells in TLOs but a much lower frequency of immunosuppressive lymphocytes. These reports suggest that it should be possible to improve and enhance the anti-cancer activity of TLOs by the conditioning and manipulation of TLOs within the cancer (Sautes-Fridman et al. 2019).

Recently, genetically engineered killer T cells that have incorporated a tumor antigen-specific antibody fragment into the T cell antigen receptor complex (CAR-T cells) have been explored as an effective option for cancer treatment (Eshhar et al. 2001; Maher et al. 2002; Kalos et al. 2011; Maude et al. 2014; Brown et al. 2016). Furthermore, a novel technology to propagate tumor antigen-specific killer T cells has been established by applying human iPS cells (Kawamoto et al. 2018; Kashima et al. 2020). However, it has sometimes been reported that the homing and long-range maintenance of iPS-derived killer T cells or CAR-T cells are not always possible, even in the presence of TILs in tumors. Hence, sufficient sites might not always be available in vivo for the maintenance, propagation and re-activation of either the endogenous TILs, CAR-T cells or iPS-derived killer T cells for an extended period of time. Therefore, a therapeutic strategy must be devised to ensure the presence of sites for immune cells, where the killer T cells can continuously self-propagate and become re-activated in parallel with the frequent and rapid mutation of tumor cells, which can alter their antigenicity to decrease the antitumor response. Recent studies have reported that microenvironments, such as the one provided by stromal cells in the cancer tissues, are suitable sites for immune tissue construction (Dieu-Nosjean et al. 2016; Sautes-Fridman et al. 2019). The formation of TLOs (not just the infiltration of killer T cells) is often clearly observed in various human cancers, such as breast, colorectal, lung, pancreatic and kidney and in malignant melanoma and lymphoma (Dieu-Nosjean et al. 2014; Joyce and Fearon 2015; Hiraoka et al. 2016; Yu et al. 2007; Figenschau et al. 2015; Kuwabara et al. 2019). The comprehensive analysis presented in these studies suggests that many clinical cases with TLO formation exhibit better prognosis than cases without TLO formation, except for cases in which the TLO predominantly comprises regulatory T cells (Hiraoka et al. 2016). This suggests that TLO formation might play a crucial role in the prevention of cancer cell proliferation by exhibiting antitumor activity.

5 Facilitation of Antitumor Effects by Active Induction of TLO Formation

The formation of TLO can be induced in tumor-bearing mice by stimulating the production of LIGHT protein (Scheu et al. 2002; Wang et al. 2009) in the LTβR-positive stromal cells within the cancer tissues. This can be accomplished by expressing the *LIGHT* gene in tumor tissues using expression vectors, such as adenoviral vectors (Yu et al. 2004, 2007). In particular, the expression of LIGHT protein in the TLOs at cancer sites promotes the differentiation, proliferation and activation of tumor-specific CD8+ killer T cells, which inhibits tumor proliferation and results in cancer regression. It should also be noted that TLO formation at tumor sites aggressively inhibits tumor proliferation, not only at the primary site but also at metastatic sites (Yu et al. 2007). The same antitumor effect of LIGHT protein has also reported by expressing it within the tumor using genetically engineered inactivated *Salmonella* encoding LIGHT (Loeffler et al. 2007). Additionally, LIGHT is also known to bind to the herpesvirus entry mediator (HVEM). Since HVEM (TNFRSF14) is a co-stimulator that promotes T cell activation, it is thought that the strong antitumor effect LIGHT is mediated not only by promoting TLO formation but also through multiple other pathways, including activation of T cells and dendritic cells (Scheu et al. 2002; Wang et al. 2009). Hence, LIGHT is essential for an effective antitumor immune response. Another study demonstrated that a potent antitumor response can be achieved by the administration of an LTβR agonistic antibody at tumor sites, which activates the LTo stromal cells through LTβR stimulation and results in TLO formation (Lukashev et al. 2006).

Blood vessels generated during neovascularization of the cancer tissues are mostly turbulent. Hence, these blood vessels are not in a functional state to allow the infiltration of immune cells, such as killer T cells, into the tumor tissues. However, TLO formation promotes remodeling of turbulent blood vessels into normal functional blood vessels, allowing infiltration of immune cells into the tumor tissues. Therefore, fusion of the LIGHT molecule with vascular targeting peptide (VTP) generates the LIGHT-VTP chimeric molecule that, through VTP, brings LIGHT into the new blood vessels in tumor tissues. This then results in TLO formation and vessel normalization in or in the vicinity of the tumor tissues (Johansson-Percival et al. 2015; Johansson-Percival et al. 2017). This strategy could result in strong antitumor responses by effectively promoting killer T cell infiltration into the tumor tissues.

Checkpoint inhibitor therapy using the anti-PD-1, anti-PD-L1 or anti-CTLA4 antibodies is receiving considerable attention as a novel immunotherapy for cancer (Brahmer et al. 2012; Topalian et al. 2012; Tang et al. 2016). This approach stably suppresses tumors for a long period by inhibiting the exhaustion of killer T cells as well as inactivating immune-suppressor T cells. However, recent studies have revealed the existence of cancers that are resistant to checkpoint inhibitor therapies such as anti-PD-L1 antibody. In addition to the anti-PD-L1 antibody, a chimeric protein (LIGHT fused to anti-EGF receptor antibody), which was produced by the fusion of the LIGHT protein to an antibody that targets a cancer cell surface antigen (EGF receptor protein), was injected into such resistant tumor tissues. This combined therapy resulted in the formation of TLOs in the tumor tissues and remarkably suppressed tumor growth (Tang et al. 2016). The CD8+ killer T cells in TLOs, which were induced by the accumulation of the LIGHT molecule at the tumor site, were activated by the anti-PD-L1 antibody and induced a strong suppression of tumor growth. While treatment with the anti-PD-L1 antibody alone or the LIGHT-EGF receptor antibody chimeric molecule alone did not result in a strong tumor regression effect, treatment with a combination of the two resulted in complete tumor regression. As stated above, many studies have indicated that TLO formation around or in cancer tissues is beneficial for suppression and regression of cancer.

6 Construction of Artificial Lymphoid Tissues (aLTs)/ Organoids

The construction of functional aLTs requires a detailed understanding of the process involved in the formation of SLOs and TLOs. aLTs have to be structurally and functionally similar to SLOs and capable of inducing antigen-specific secondary immune responses. Additionally, aLTs must mimic the dynamic kinetics of immune cells inside and outside the immune tissue. Although the formation of TLOs during inflammation is usually transient, aLTs must remain structurally and functionally stable for an extended period in the body. Furthermore, aLTs must be constructed on a scaffold structure so that their extraction can be smoothly executed for transplantation into the recipients. Hence, it is important to select a suitable material for the scaffold that has high biocompatibility and excellent operability for aLT construction.

The artificial construction of immune tissues has been proposed as a candidate for treating immunological diseases (Cupedo et al. 2012). We have first succeeded in constructing artificial lymph node-like lymphoid tissues (aLTs) in a mouse (Suematsu and Watanabe 2004, Lenti et al. 2019). The aLT tissues were lymphoid tissues similar to the secondary lymph nodes, exhibited a structure similar to the natural lymph node and possessed potent immune activity. We have demonstrated that this particular aLT not only produces antigen-specific antibodies but also facilitates enhanced antitumor responses (Okamoto et al. 2007; Kobayashi et al. 2011;

Kobayashi and Watanabe 2010). We have also shown the feasibility of constructing aLTs without using any stromal cells as described below. We have likewise succeeded in regenerating and reconstructing another secondary immune organ, the spleen of mouse, which consisted of both white pulp as well as red pulp, similar to the normal spleen. The re-constructed spleen could mount antigen-specific immune responses (Tan and Watanabe 2014, 2017). In the following sections, we present the method for artificial construction of lymphoid tissue in mice, which exhibits promising immune-stimulatory activity and may be useful for the treatment of severe infection, immunodeficiency and cancers (Suematsu and Watanabe 2004; Okamoto et al. 2007).

7 Artificial Construction of aLTs by Applying Stromal Cells in Mice

In order to construct aLTs with a structure similar to the natural secondary lymph node in which immune cells such as B cells, T cells and dendritic cells are efficiently accumulated and arranged, we first established a mouse lymphoid tissue-specific stromal cells expressing LTβR and TNFαR (Suematsu and Watanabe 2004). As the three-dimensional scaffold, biocompatible high polymer materials, such as collagen sponges or collagen sheets, were applied. The mixtures of mouse lymphoid tissue-specific stromal cells and bone marrow-derived dendritic cells were adsorbed into 2–3 mm square collagen sponges. Then, the collagen sponges containing stromal cells and dendritic cells were implanted into the renal subcapsular space of naïve (BALB/c) mice (Fig. 1a). After 2–3 weeks, a lymphocyte mass several millimeters in size was formed. The structure of the lymphocyte mass was similar to that of natural lymph nodes, with clearly defined T and B cell zones. The artificially constructed lymphoid tissues had the following characteristics (Suematsu and Watanabe 2004; Okamoto et al. 2007). (1) They comprised of dendritic cells, T cells and B cells. (2) The T and B cell zones were clearly segregated. (3) They had a germinal center (GC), and active B cell proliferation was observed in response to antigen stimulation, similar to the B cells in GCs of normal lymphoid tissue. (4) Formation of FDC networks in the B cell zone was evident. (5) Formation of HEVs with a well-developed capillary plexus and small functional veins was evident. (6) Formation of many functional lymph and blood capillaries around the aLT was detected. (7) Two types of stromal cell networks supporting the lymph node structure were clearly formed; the FDC network in B cell follicles and the FRC network, minutely stretched around the T cell zone. It is important to note that the structures of the FRC and FDC networks appeared to be formed prior to the influx of T and B cells into the aLT. (8) The aLT structure was stable for an extended period in the mouse. (9) The aLTs were easily extracted and transplanted into other individuals. (10) As we discuss in the following section, aLT transplantation may restore the immunocompetence of the recipient. However, it should be noted that autoantibody production was not detected within aLT or in the aLT-implanted

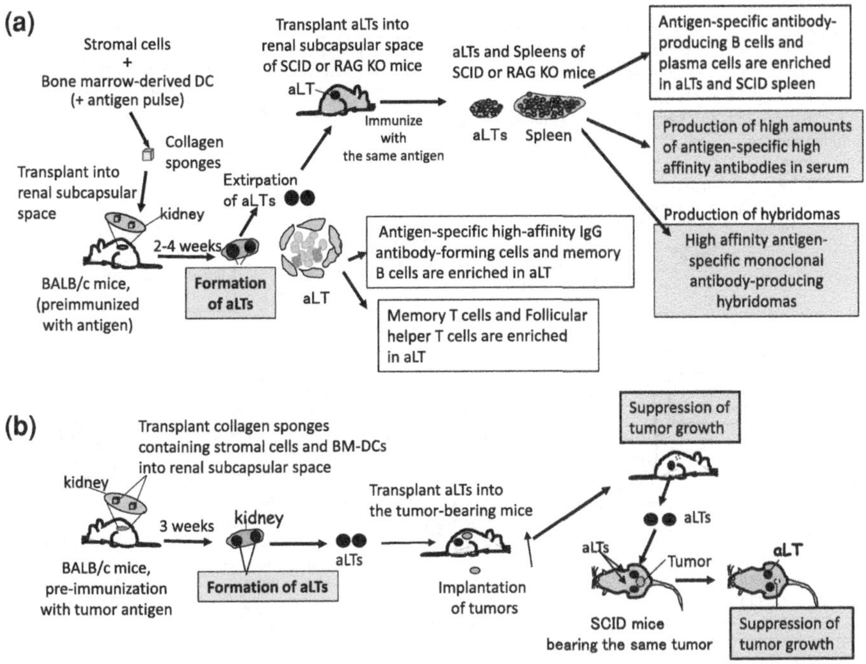

Fig. 1 Artificial construction of lymphoid tissues (aLTs)/organoids in mouse and human. **a** Antigen-specific high-affinity antibody-producing cells are effectively enriched in the aLTs and the spleen of aLT-transplanted SCID mice upon antigen re-immunization. **b** The aLTs, constructed in mice and pre-sensitized with tumor antigen, exhibit a strong suppressive activity against the tumor growth when transplanted into tumor-bearing recipients. **c** Construction of aLTs in previously tumor-bearing mice that had undergone surgical eradication of the tumor effectively suppressed tumor relapse. **d** Strategy to construct human-type aLTs by combining human stromal cell spheroids and human peripheral blood mononuclear cells (PMBCs)

individuals. (11) The aLT should be categorized as a TLO as it does not form a cortex/medulla structure and it does not have capsular structure to cover the external surface.

8 Immune Function of the aLTs/Organoids

The aLT constructed in naïve (BALB/c) mice could induce a much stronger antigen-specific responses than that in the spleen of recipient mice (Okamoto et al. 2007). After construction of aLTs in the renal subcapsular space of mice that had been pre-sensitized with antigen, the aLTs were extracted and transplanted into the renal subcapsular space of naïve mice. The mice carrying aLT were then immunized with the same antigen. A strong secondary immune response was immediately induced only in the aLT but not in the recipient spleen. The antigen-specific

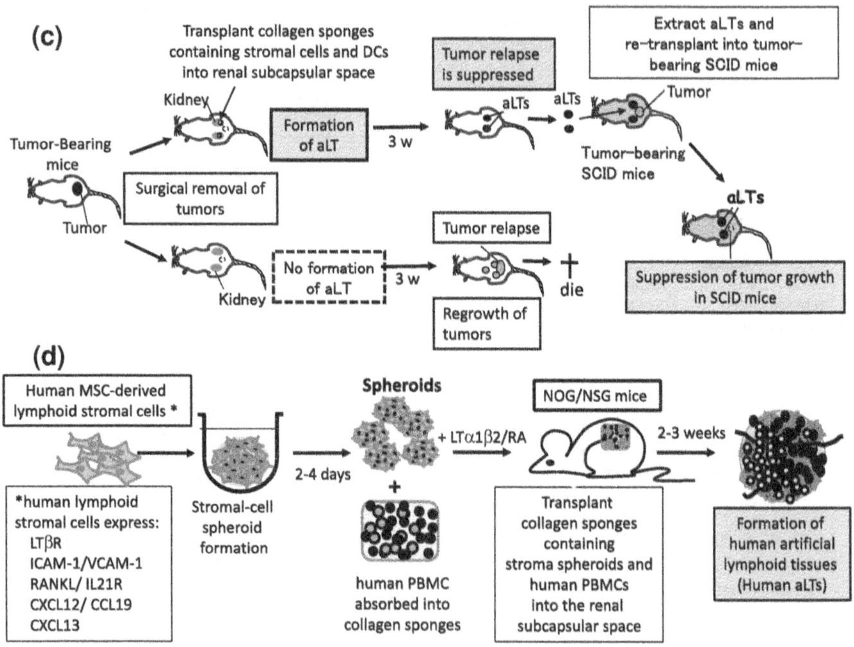

Fig. 1 (continued)

high-affinity IgG antibody-forming cells (AFCs) emerged with high efficiency in the aLTs, and large quantities of antigen-specific high-affinity IgG antibodies were detected in the recipient mouse serum. However, only IgM AFCs were detected in the spleen or lymph nodes of the recipient mice due to the primary immune response (Okamoto et al. 2007). The aLTs were also transplanted into renal subcapsular space of SCID mice (an immunodeficient mouse model lacking B and T cells), and then, the SCID mice were immunized with the same antigen. A strong secondary immune response was immediately induced in the transplanted aLT, and antigen-specific high-affinity IgG AFCs emerged in large quantities in the aLT. In parallel, antigen-specific high-affinity AFCs gradually emerged in the spleen as well as in the bone marrow of the aLT-implanted SCID mice (Fig. 1a). The numbers of high-affinity AFCs produced in the spleen of the aLT-implanted SCID mice were 10–50 times higher than in the spleen of the aLT-implanted normal naïve mice (Okamoto et al. 2007). A strong secondary immune response was induced locally at the aLTs in the aLT-implanted normal as well as SCID mice. Surprisingly, in the aLT-implanted immunodeficient SCID mice, the AFCs quickly moved to a "vacant" lymphoid tissues (such as the spleen) from the aLTs and rapidly proliferated and differentiated into antigen-specific ultra-high-affinity AFCs. It should be possible to establish an antigen-specific ultra-high-affinity antibody by constructing antibody-producing hybridomas using such spleen cells. These results suggest the possibility of inducing an effective and strong immune response by transplanting the aLT into individuals with severe infections or aberrant immune function to induce an effective and strong immune response. The memory type helper T cells and memory B cells were also concentrated selectively at high frequency in the aLT.

We then constructed the aLT by implanting stromal cells together with bone marrow-derived dendritic cells (BM-DCs), which were enclosed in a collagen sponge, into the renal subcapsular space of normal mice that had been pre-immunized with tumor-specific antigen (Fig. 1b). The aLT was then transplanted into the renal subcapsular space of tumor-bearing SCID mice that had been pre-transplanted with the same tumor. We observed that tumor proliferation was suppressed in these mice (Kobayashi et al. 2011; Kobayashi and Watanabe 2010). Various INFγ-producing cells were observed in the aLT, mostly comprised of CD8 + killer T cells, CD4+ helper T cells and NK cells, which probably promoted a potent antitumor response. This finding indicates that the aLT is capable of strongly inducing not only a humoral but also a cell-mediated immune response.

In another study, tumor cells were transplanted under the skin of naïve mice and allowed to grow to form a tumor mass. The tumors were then surgically extirpated when they reached a certain size, and the mice were then divided into two groups. An aLT was constructed in one group but not the other group. After three to four weeks, all mice without an aLT construction died from cancer recurrence. However, the mice with the aLT all survived without cancer recurrence or with only a small recurrence under strong suppression (Fig. 1c). Taken together, the aLT could induce a strong humoral as well as a strong cell-mediated secondary immune reaction.

9 Construction of aLTs Without Using Stromal Cells

We also constructed aLT/organoid in vivo by using only humoral factors without the use of stromal cells (Kobayashi and Watanabe 2016). Mixtures of the chemokines, CCL19, CCL21, CXCL12 and CXCL13 that promote immune cell migration, as well as the recombinant LTα1β2 protein, were absorbed in a sustained-release gel (Medgel). The gels were enclosed in the collagen sponge to which the adhesion molecule, recombinant VCAM1, was attached. Then, the collagen sponge was transplanted into the renal subcapsular space of naïve (BALB/c) mice. The formation of lymphoid tissue comprising T and B cell clusters was observed in the graft with excellent reproducibility (Kobayashi and Watanabe 2016; Kobayashi et al. 2016). The formation of the FRC stromal cell network, the distribution of various dendritic cells in the T cell cluster and the local existence of FDCs in the B cell cluster was all evident. The aLT was constructed in BALB/c mice pre-immunized with the antigen by using a sustained-release gel containing chemokines. The aLT was then removed and re-implanted into the renal subcapsular space of immunodeficient SCID mice, which were subjected to secondary immunization with the same antigen. Highly efficient induction of antigen-specific high-affinity IgG AFCs was observed in the aLT as well as in the spleen of the recipient SCID mice, similar to that observed using aLT tissue constructed with stromal cells. These results demonstrate that immunologically functioning and stable aLTs can be constructed in mice without stromal cells, but by using only soluble factors in combination with various chemokines and that the efficiency of the immune function of the aLT was similar to that of the aLT constructed with stromal cells.

10 Construction of Human-Type aLT/Organoid

Generation of functional human-type artificial lymphoid tissues/organoids should be beneficial for clinical application as immunotherapy against severe infection, cancer, autoimmune diseases or immunodeficiency caused by aging and also for investigation of the functions of human lymphoid tissues. As stated above, the immunologically active artificially lymphoid tissues are able to be stably generated in mice by applying either lymphoid stromal cell-embedded biocompatible scaffold or the chemokines/lymphokines gel-embedded scaffold (Suematsu and Watanabe 2004; Kobayashi and Watanabe 2016). Considering the various studies on the SLOs and TLOs in human (Ruddle 2014; Dieu-Nosjean et al. 2014; Turley et al. 2015; Johasson et al. 2016; Tschenig et al. 2018; Fennema et al. 2013), it is clear that the appropriate stromal cells are required also for the formation of human-type artificial lymphoid tissues/organs. As a candidate human lymphoid stromal cell, we established a stromal cell clone from the commercially available human bone marrow-derived mesenchymal stem cell lines (RIKEN Institute, Japan). It expresses LTβ receptors and adhesion molecules such as VCAM-1, ICAM-1 and MadCAM-1 on the cell surface. The cloned stromal cell increased the expression of various lymphoid chemokines, such as CXCL13, CCL19, CCL21 CXCL12 and adhesion molecules in response to stimulation with retinoic acid and LTα1β2 ligand (Fig. 1d). Next, the stromal cells were subjected to the spheroid formation (Yin et al. 2017) to establish a three-dimensional structure. The stromal cell spheroids were then placed on collagen sponges onto which human peripheral blood mononuclear cells (PBMC) had been pre-absorbed. Formation of blood vessel and lymphatic tracts in the spheroids enhanced the influx and accumulation of human PBMCs into the stromal spheroids. The collagen sponges carrying stromal spheroids and human PBMCs were then transplanted into the renal subcapsular space of immunodeficient NOG or SCID mice. After a few weeks, tertiary lymphoid tissues containing clusters of the human T and B cells and scattered human DC cells were formed (Fig. 1d). Immunodeficient mice carrying the human-type artificially made lymphoid tissues (human-type aLTs) were then immunized with viral antigens. Strong B cell proliferation and germinal center formation and appearance of antigen-specific antibody-producing cells were observed in the aLTs as a result of the immunization. Human IFNγ production was also detected in the T cells of mice harboring the aLTs. These data indicate that immunologically functional human-type aLTs could be generated by directly applying human PBMC into the human stromal spheroids. We are now planning to induce tumor-killing T cells by combining human aLTs with PDX (patient-derived xenografts) mice carrying human cancers.

11 In Summary

We hope that aLT construction will be developed further and will be applied in the future, not only as a novel clinical device but also as a new experimental system for basic immunology research. Additionally, we believe that artificial restoration/construction and use of immune tissues other than lymph node tissue, such as the thymus, bone marrow or intestinal mucosa, will be challenging but important.

Our endeavor to artificially construct the immunologically functioning-lymphoid tissue/organoid has just begun. We sincerely hope that more researchers will contribute to the field of the reconstructive immunology or immune tissue engineering in the future.

References

Ager A (2017) High Endothelial Venules and other blood vessels: critical regulators of lymphoid organ development and function. Front Immunol 8:45. https://doi.org/10.3389/fimmu.2017.00045

Bar-Ephraim YE, Konijn T, Gonuitas M, Mebius RE, Reijimers M (2016) A reproducible method for isolation and in vitro culture of functional human lymphoid stromal cells from tonsils. PLoS ONE 11(12)e0167555. https://doi.org/10.1371/journal.pone.0167555

Brahmer JR, Tykod SS, Chow LQ, Hwu WJ, Topalian SL, Hwu P et al (2012) Safety and activity of anti-PD-L1 antibody in patients with advanced cancer. N Engl J Med 365:2455–2465. https://doi.org/10.1056/NEJMoa1200694

Brown CE, Allzadeh D, Starr R, Weng L, Wagner JR et al (2016) Regression of glioblastoma after chimeric antigen receptor T-cell therapy. New Engl J Med 375:2561–2016. https://doi.org/10.1056/NEJMoa1610497

Buckley CD, Barone F, Nayar S, Benezech C, Caamano L (2015) Stromal cells in chronic inflammation and tertiary lymphoid organ formation. Annu Rev Immunol 33:715–745. https://doi.org/10.1146/annrev-immunol-032713-120252

Carragher DM, Rangel-Moreno J, Randall TD (2008) Ectopic lymphoid tissues and local immunity. Semin Immunol 20:26–42. https://doi.org/10.1069/j-smim.2007.12.004

Chai Q, Onder L, Scandella E, Gil-Cruz C, Perez-Shibayama C, Cupovic J et al (2013) Maturation of lymph node fibroblastic reticular cells from myofibroblastic precursors is critical for antiviral immunity. Immunity 38:1013–1024. https://doi.org/10.1016/j.immuni.2013.03.012

Colbeck EJ, Ager A, Gallimore A, Jones GW (2017) Tertiary lymphoid structures in cancer: drivers of antitumor immunity, immunosuppression, or Bystander sentinels in diseases? Front Immunol 8:1830. https://doi.org/10.3389/fimmu.2017/01830

Couillault C, Germain C, Dubois B, Kaplon H (2018) Identification of tertiary lymphoid structure-associated follicular helper T cells in human tumors and tissues. Methods Mol Biol 1845:205–222. https://doi.org/10.1007/978-1-4939-8709-2_12

Crotty S (2011) Follicular helper CD4 T cells (T-FH). Annu Rev Immunol 29:621–663. https://doi.org/10.1146/annurev-immunol-031210-101400

Cupedo T, Stroock A, Coles M (2012) Application of tissue engineering to the immune system: development of artificial lymph nodes. Front Immunol 3:343. https://doi.org/10.3389/fimmu.2012.00343

de Leur K, Clahsen-van Groningen MC, van den Bosch TPP, der Graav GN, Hesselink DA, Samsom JN, Baan CC, Boer K (2018) Characterization of ectopic lymphoid structures in different types of acute renal allograft rejection. Clic Exp Immunol 192:224–232. https://doi.org/10.1111/cei.13099

Dieu-Nosjean MC, Goc J, Giraldo NA, Sautes-Fridman C, Fridman WH (2014) Tertiary lymphoid structures in cancer and beyond. Trends Immunol 35:571–580. https://doi.org/10.1016/j.it.2014.09.006

Dieu-Nosjean MC, Giraldo NA, Kaplon H, Germain C, Fridman WH, Sautes-Fridman C (2016) Tertiary lymphoid structures, drivers of the anti-tumor responses in human cancers. Immunol Rev 271:260–275. https://doi.org/10.1111/imr.12405

Dorraji SE, Hovd A-MK, Kanapathippillai P, Bakland G, Eilertsen G, Figenschau SL, Fenton KA (2018) Mesenchymal stem cells and T cells in the formation of tertiary lymphoid structures in lupus nephritis. Sci Rep 8:7861. https://doi.org/10.1038/s41598-018-26265-z

Drayton DL, Liao S, Mounzer RH, Ruddle NH (2006) Lymphoid organ development: from ontogeny to neogenesis. Nat Immunol 7:344–353. https://doi.org/10.1038/ni1330

Engelhard VH, Rodriguez AB, Mauldin IS, Woods AN, Peske JD, Slingluff CL (2018) Immune cell infiltration and tertiary lymphoid structures as determinants of antitumor immunity. J Immunol 200:432–442. https://doi.org/10.4049/jimmunol.1701269

Eshhar Z, Waks T, Bendavid A, Schindler DG (2001) Functional expression of chimeric receptor genes in human T cells. J Immunol Methods 248:67–76. https://doi.org/10.1016/s0022-1759(00)00343-4

Fennema E1, Rivron N, Rouwkema J, van Blitterswijk C, de Boer J (2013) Spheroid culture as a tool for creating 3D complex tissues. Trends Biotechnol 31:108–115. https://doi.org/10.1016/j.tibtech.2012.12.003

Figenschau SL, Fismen S. Fenton KA, Fenton C, Mortensen ES (2015) Tertiary lymphoid structures are associated with higher tumor grade in primary operable breast cancer patients. BMC Cancer (101). https://doi.org/10.1186/s12885-015-1116-1

Fletcher AL, Malhotra D, Acton SE, Lukacs-Kornek V, Bellemare-Pelletier A, Curry M et al (2011) Reproducible isolation of lymph node stromal cells reveals site-dependent differences in fibroblastic reticular cells. Front Immunol 2:35. https://doi.org/10.3389/fimmu.2011.00035

Fonseca VR, Romao VC, Aqua-Doca A, Santos M, Lopez-Presa D, Ferreira AC, Fonseca JE, Graca L (2018) The ratio of blood T follicular regulatory cells to T follicular helper cells marks ectopic lymphoid structure formation while activated follicular helper T cells indicate disease activity in primary Sjogren syndrome. Arthritis Rheumatol 70:774–784. https://doi.org/10.1002/art.40424

Fu YX, Chaplin DD (1999) Development and maturation of secondary lymphoid tissues. Annu Rev Immunol 17:399–433. https://doi.org/10.1146/annurev.Immunol.17.1.399

Gommerman JL, Browning JL (2003) Lymphotoxin/LIGHT, lymphoid microenvironments and disease. Nat Rev Immunol 3:642–655. https://doi.org/10.1038/nri1151

Grabner R, Lotzer K, Dopping S, Hildner M, Radke D, Beer M et al (2009) Lymphotoxin beta receptor signaling promotes tertiary lymphoid organogenesis in the aorta adventitia of aged ApoE-/- mice. J Exp Med 206:233–248. https://doi.org/10.1084/jem.20080752

Halle S, Dujardin HC, Bakocevic N, Fleige H, Danzer H, Willenzon S, Suezer Y, Hammerling G, Garbi N, Sutter N, Worbs T, Forster R (2009) Induced bronchus-associated lymphoid tissue serves as a general priming site for T cells and is maintained by dendritic cells. J Exp Med 206:2593–2601. https://doi.org/10.1084/jem.20091472

Hiraoka N, Ino Y, Yamazaki-Itoh R (2016) Tertiary lymphoid organs in cancer tissues. Front Immunol 7:244. eCollection 2016 https://doi.org/10.3389/fimmu.2016.00244

Hu D, Mohanta SK, Yin C, Peng L, Ma Z, Srikakulapu P et al (2015) Artery tertiary lymphoid organs control aorta immunity and protect against atherosclerosis via vascular Smooth muscle cell lymphotoxin beta Receptors. Immunity 42:1100–1115. https://doi.org/10.1016/j.immuni.2015.05.015

Hughes CE, Benson RA, Bedaj M, Maffia P (2016) Antigen-presenting cells and antigen presentation in tertiary lymphoid organs. Front Immunol 7:481. eCollection 2016. https://doi.org/10.3389/fimmu.2016.00481

Johansson-Percival A et al (2015) Intratumoral LIGHT restores pericyte contractile properties and vessel integrity. Cell Rep 13:2687–2698. https://doi.org/10.1016/j.celrep.2015.12.004

Johansson-Percival A, He B, Li ZJ, Kjellen A, Russel K, Li Ji, Larma I, Ganss R (2017) De novo induction of intratumoral lymphoid structures and vessel normalization enhances immunotherapy in resistant tumors. Nat Immunol 18:1207–1217. https://doi.org/10.1038/ni.3836

Johasson A, Hamzah J, Ganss R (2016) More than a scaffold: stromal modification of tumor immunity. Biochim Biophys Acta 1865:3–13. https://doi.org/10.1016/j.bbcan.2015.06.001

Jones GW, Jones SA (2016) Ectopic lymphoid follicles: inducible centers for generating antigen-specific immune responses within tissues. Immunology 147:141–151. https://doi.org/10.1111/imm.12554

Jones E, Gallimore A, Ager A (2018) Defining high endothelial venules and tertiary lymphoid structures in cancer. Methods Mol Biol 1845:99–118. https://doi.org/10.1007/978-1-4939-8709-2_7

Joyce JA, Fearon DT (2015) T cell exclusion, immune privilege, and the tumor microenvironment. Science 348:74–80. https://doi.org/10.1126/science.aaas6204

Kalos M, Levine BL, Porter DL, Katz S, Grupp SA, Bagg A, June CH (2011) T cells with chimeric antigen receptors have potent antitumor effects and can establish memory in patients with advanced leukemia. Sci Transl Med 3(95):95ra73. https://doi.org/10.1126/scitransmed.300284

Kashima S, Kawamoto H, et al (2020) Cytotoxic T lymphocytes regenerated from iPS cells have therapeutic efficiency in a patient-derived xenograft solid tumor model. iScience 23:100998. https://doi.org/1.1016/j.isci.2020.100998

Kawamoto H, Masuda K, Nagao S, Maeda T (2018) Cloning and expansion of antigen-specific T cells using iPS cell technology: development of “off-the-shelf” T cells for the use in allogeneic transfusion settings. Int J Hematol 107:271–277. https://doi.org/10.1007/s12185-018-2399-1

Kobayashi Y, Watanabe T (2010) Synthesis of artificial lymphoid tissue with immunological function. Trends Immunol 31:422–428. https://doi.org/10.1016/j.it.2010.09.002

Kobayashi Y, Watanabe T (2016) Gel-trapped lymph-organogenic chemokines trigger artificial tertiary lymphoid organs and mount adaptive immune responses in vivo. Front Immunol 7:316. https://doi.org/10.3389/fimmu.2016.00316

Kobayashi Y, Kato K, Watanabe T (2011) Synthesis of functional artificial lymphoid tissues. Discov Med 12:351–362 PMID:22031672

Kobayashi S, Murata K, Shibuya H, Morita M, Ishikawa M, Furu M et al (2013) A distinct human CD4+ T cell subset that secretes CXCL13 in rheumatoid synovium. Arthritis Rheum 65:3063–3072. https://doi.org/10.1002/art.38173

Kobayashi Y, Kato K, Nakamura M, Watanabe T (2016) Synthresis of functional tertiary lymphoid organs. In: Watanabe T, Takahama Y (eds) Synthetic immunology. Springer Nature, pp 151–169. https://doi.org/10.1007/978-4-431-56027-2_7

Koning JJ, Mebius RE (2012) Interdependence of stromal and immune cells for lymph node function. Trends Immunol 33:264–270. https://doi.org/10.1016/j.it.2011.10.006

Kroeger DR, Milne K, Nelson BH (2016) Tumor-infiltrating plasma cells are associated with tertiary lymphoid structures, cytolytic T-cell responses, and superior prognosis in ovarian cancer. Clin Cancer Res 22:3005–3015. https://doi.org/10.1158/1078-0432.CCR-15-2762

Kuwabara S, Tsuchikawa T, Nalamura T, Hatanaka Y, Hatanaka K, Sasaki K, Ono M, Umemoto K, Suzuki T, Sato O, Hane Y, Nakanishi Y, Asano T, Ebihara Y, Kurashima Y, Noji T, Murakami S, Okamura K, Shichinohe T, Hirano S (2019) Prognostic relevance of tertiary organs following neoadjuvant chemoradiotherapy in pancreatic ductal adenocarcinoma. Cancer Sci 110:1853–1862. https://doi.org/10.1111/cas.14023

Lenti E, Brendolan A, et al (2019) Therapeutic regeneration of lymphatic and immune cell function upon lympho organoid transplantation. Stem Cell Reports 12:1260-1268. https://doi.org/10.1016/j.stemcr.2019/04.021

Lin L1, Hu X1, Zhang H1, Hu H1 (2019) Tertiary lymphoid organs in cancer immunology: mechanisms and the new strategy for immunotherapy. Front Immunol 10:1398. https://doi.org/10.3389/fimmu.2019.0139887

Loeffler M, Le'nergrate G, Krajewska M, Reel JC (2007) Attenuated *Salmonella engineered* to produce human cytokine LIGHT inhibit tumor growth. Proc Nat Acad Sci USA 104:12879–12883. https://doi.org/10.1073/pnas.0701959104

Lotzer K, Dopping S, Connert S, Grabner R, Spanbroek R, Lemser B et al (2010) Mouse aorta smooth muscle cells differentiate into lymphoid tissue organizer-like cells on combined tumor necrosis factor receptor-1/lymphotoxin beta-receptor NF-kappaB signaling. Arterioscler Thromb Vasc Biol 30:395–402. https://doi.org/10.1161/ATVBAHA.109.191395

Lukashev M, Lepage D, Wilson C, Bally V, Garber E, Lukashin A et al (2006) Targeting the lymphotoxin-(beta) receptor with agonist antibodies as a potential cancer therapy. Cancer Res 66:9617–9624. https://doi.org/10.1158/0008-5472.CAN-06-0217

Maher J, Brentjens RJ, Gunset G, Riviere I, Sadelain M (2002) Human T-lymphocye cytotoxicity and proliferation directed by a single chimeric TCRzeta/CD28 receptor. Nat Biotechnol 20:70–75. https://doi.org/10.1038/nbt0102-70

Maude SL, Frey N, Shaw PA, Aplenc R, Barrett DM et al (2014) Chimeric antigen receptor T cells for sustained remissions in leukemia. N Engl J Med 371:1507–1517. https://doi.org/10.1056/nejmoa1407222

Mebius RE (2003) Organogenesis of lymphoid tissues. Nat Rev Immunol 63:292–303. https://doi.org/10.1038/nri1054

Nasr IW, Reel M, Oberbarnscheidt MH, Mounzer RH, Baddoura FK, Ruddle NH et al (2007) Tertiary lymphoid tissues generate effector and memory T cells that lead to allograft rejection. Am J Transplant 7:1071–1079. https://doi.org/10.1111/j.1600-6143.2007.01756.x

Neyt K, Perros F, GeurtsvanKessel CH, Hammad H, Lambrecht BN (2012) Tertiary lymphoid organs in infection and autoimmunity. Trends Immunol 33:297–305. https://doi.org/10.1016/j.it.2012.04.006

Okamoto N, Chihara R, Shimizu C, Nishimoto S, Watanabe T (2007) Artificial lymph nodes induce potent secondary immune responses in naive and immunodeficient mice. J Clin Invest 117:997–1007. https://doi.org/10.1172/jci30379

Randall TD, Carragher DM, Rangel-Moreno J (2008) Development of secondary lymphoid organs. Annu Rev Immunol 26:627–650. https://doi.org/10.1146/annurev.immunol.26.021607.090257

Roco JA, Mesin L, Binder CB, Nefzger C, Gonzalez-Figueroa P, Canete PF, Ellyard J, Shen Q, Robert PA, Cappelo J, Vohra H, Zhang Y, Nowosad CR, Schiepers A, Corcoran LM, Toellner K-M, Polo JM, Meyer-Hermann M, Victora GD, Vinuesa CG (2019) Class-switch recombination occurs infrequently in germinal centers. Immunity 51:337–350. https://doi.org/10.1016/j.immuni.2019.07.001

Roozendaal R, Mebius RE (2011) Stromal cell-immune cell interactions. Annu Rev Immunol 29:23–437. https://doi.org/10.1146/annurev-immunol-031210-101357

Rosenberg SA, Dudley ME (1990) Adoptive cell therapy for the treatment of patients with metastatic melanoma. Curr Opin Immunol 21:233–240. https://doi.org/10.1016/j.coi2009.03.002

Ruddle NH (2014) Lymphatic vessels and tertiary lymphoid organs. J Clin Invest 124:953–959. https://doi.org/10.1172/JCI71611

Salomonsson S, Jonsson MV, Skarstein K, Brokstad KA, Hjelmstrom P, Wahren- Herlenius M et al (2003) Cellular basis of ectopic germinal center formation and autoantibody production in the target organ of patients with Sjogren's syndrome. Arthritis Rheum 48:3187–3201. https://doi.org/10.1002/art.11311

Sautes-Fridman C, Petitprez F, Calderaro J, Fridman WH (2019) Tertiary lymphoid structures in the era of cancer immunotherapy. Nature Rev Cancer 19:307–325. https://doi.org/10.1038/s41568-019-0144-6

Scheu S, Alferink J, Potzel T, Barchet W, Kalinke U, Pfeffer K (2002) Targeted disruption of LIGHT causes defects in costimulatory T cell activation and reveals cooperation with lymphotoxinbeta in mesenteric lymph node genesis. J Exp Med 195:1613–1624. https://doi.org/10.1084/jem.20020215

Shi K, Hayashida K, Kaneko M, Hashimoto J, Tomita T, Lipsky PE et al (2001) Lymphoid chemokine B cell-attracting chemokine-1 (CXCL13) is expressed in germinal center of ectopic lymphoid follicles within the synovium of chronic arthritis patients. J Immunol 166:650–655. https://doi.org/10.4049/jimmunol.166.1.650

Sicard A, Chen CC, Morelon E, Thaunat O (2016) Alloimmune-induced intragraft lymphoid neogenesis promotes B-cell tolerance breakdown that accelerates chronic rejection. Curr Opin Organ Transplant 21:368–373. https://doi.org/10.1097/MOT.0000000000000329

Silina K, Burkhardt C, Casanova R, Solterman A, van den Broek M (2018) A quatitative pathology approach to analyze the development of human cancer-associated tertiary lymphoid structures. Methods Mol Biol 1845:71–86. https://doi.org/10.1007/978-1-4939-8709-2_5

Solinas C, Garaud S, De Silva P, Boisson A, Van den Eynden G, de Wind A, Bisso P, Rodrigues-Vitoria J, Richard F, Migllori E, Noel G, Duvillier H, Craclun L, Vevs I, Awada A, Detours V, Larsimont D, Piccart-Gebhart M, Willard-Gallo K (2017) Immune checkpoint molecules on tumor-infiltrating lymphocytes and their association with tertiary lymphoid structures in human breast cancer. Front Immunol 8:1412. https://doi.org/10.3389/fimmu.2017.01412

Stott DI, Hiepe F, Hummel M, Steinhauser G, Berek C (1998) Antigen driven clonal proliferation of B cells within the target tissue of an autoimmune disease. The salivary glands of patients with Sjogren's syndrome. J Clin Invest 102:938–946. https://doi.org/10.1172/jci3234

Suematsu S, Watanabe T (2004) Generation of a synthetic lymphoid tissue-like organoid in mice. Nat Biotechnol 22:1539–1545. https://doi.org/10.1038/nbt1039

Takemura S, Braun A, Crowson C, Kurtin PJ, Cofield RH, O'Fallon WM, Goronzy JJ, Weyand CM (2001) Lymphoid neogenesis in rheumatoid synovitis. J. Immunol 167:1072–1080. https://doi.org/10.4049/jimmunol.167.2.1072

Tan JKH, Watanabe T (2014) Murine spleen tissue regeneration from neonatal spleen capsulerequires lymphotoxin priming of stromal cells. J Immunol 193:1194–1203. https://doi.org/10.4049/jimmunol.1302115

Tan JKH, Watanabe T (2017) Stromal cell subsets directing neonatal spleen regeneration. Sci Rep 7:40401. https://doi.org/10.1038/srep40401

Tang H, Wang Y, Chlewicki LK, Zhang Y, Guo J, Liang W, Wang J, Wang X, Fu YX (2016) Facilitating T cell infiltration in tumor microenvironment overcomes resistance to PD-L1 blockade. Cancer Cell 29:285–296. https://doi.org/10.1016/j.ccell.2016.02.004

Thaunat O, Graff-Dubois S, Brouard S, Gautreau C, Varthaman A, Fabien N et al (2010a) Immune responses elicited in tertiary lymphoid tissues display distinctive features. PLoS ONE 30: e11398. https://doi.org/10.1371/journal.pone.0011398

Thaunat O, Patey N, Caligiuri G, Gautreau C, Mamani-Matsuda M, Mekki Y et al (2010b) Chronic rejection triggers the development of an aggressive intragraft immune response through recapitulation of lymphoid organogenesis. J Immunol 185:717–728. https://doi.org/10.4049/jimmunol.0903589

Thomas JA, Willcox HN, Newsom-Davis J (1982) Immunohistological studies of the thymus in myasthenia gravis. Correlation with clinical state and thymocyte culture responses. J Neuroimmunol 3:319–335. https://doi.org/10.1016/0165-5728(82)90035-2

Topalian SL, Hod FS, Brahmer JR, Gettinger SN, Smith DC et al (2012) Safety, activity, and immune correlates of anti-PD1 antibody in cancer. N Engl J Med 366:2443–2454. https://doi.org/10.1056/NEJMoa1200690

Tschenig T, Pabst R, Prenzier F, Rittinghausen S, Braun A, Hohlfeld JM (2018) Isolated aggregates of lymphoid cells in the inner bronchial wall in asthma patients. Cell Tissue Res 374:423–425. https://doi.org/10.1007/s00441-018-2936-5

Turley SJ, Cremasco V, Astarita JL (2015) Immunological hallmarks of stromal cells in the tumor microenvironment. Nature Rev Immunol 15:669–682. https://doi.org/10.1038/nri.3902

Wang PL, Zhu M, Miller M, Fu YX (2009) Immunoregulation by tumor necrosis factor superfamily member LIGHT. Immunol Rev 229:232–243. https://doi.org/10.1111/j.1600-065X.2009.00762,x

Weinstein AM, Girald NA, Petitprez F, Julie C, Lacroix L, Peschaud F, Emile JF, Marisa L, Fridman WH, Storkus WJ, Sautes-Fridman C (2019) Association of IL-36γ with tertiary lymphoid structures and inflammatory immune infiltrates in human colorectal cancer. Cancer Immunol Immunother 68:109–120. https://doi.org/10.1007/s00262-018-2259-0

Weiss JM, Cufi P, Le Panse R, Berrih-Aknin S (2013) The thymus in autoimmune Myasthenia Gravis: Paradigm for a tertiary lymphoid organ. Rev Neurol (Paris) 169:640–649. https://doi.org/10.1016/j.neurol.2013.02.005

Yin C, Mohanta S, Maffla P, Habenicht AJR (2017) Editorial: tertiary lymphoid organs (TLOs): powerhouses of disease immunity. Front Immunol 8(228). https://doi.org/10.3389/fimmu.2017.00228

Young CL, Adamson TC, 3rd, Vaughan JK, Fox RI (1984) Immunohistologic characterization of synovial membrane lymphocytes in rheumatoid arthritis. Arthritis Rheum 27: 32–39. https://doi.org/10.1002/art.1780270106

Yu P, Lee Y, Liu W, Chin RK, Wang J, Wang Y, Schietinger A, Philip M, Schreiber H, Fu YX (2004) Priming of naïve T cells inside tumors leads to eradication of established tumors. Nat Immunol 5:141–149. https://doi.org/10.1038/ni1029

Yu P, Lee Y, Wang Y, Liu X, Auh S, Gajeski TF, Schreiber H, You AZ, Kayor C, Wang X, Fu XY (2007) Targeing the primary tumors to generate CTL for the effective eradication of spontaneous metastasis. J Immunol 179:1960–1968. https://doi.org/10.4049/jimmunol.179.3.1960